Energy Harvesting: Principles, Modeling and Applications

Energy Harvesting: Principles, Modeling and Applications

Edited by
Michael Stock

Larsen & Keller
www.larsen-keller.com

Energy Harvesting: Principles, Modeling and Applications
Edited by Michael Stock
ISBN: 978-1-63549-684-0 (Hardback)

Larsen & Keller

Published by Larsen and Keller Education,
5 Penn Plaza,
19th Floor,
New York, NY 10001, USA

Cataloging-in-Publication Data

Energy harvesting : principles, modeling and applications / edited by Michael Stock.
    p. cm.
Includes bibliographical references and index.
ISBN 978-1-63549-684-0
1. Energy harvesting. 2. Power resources. I. Stock, Michael.
TK2897 .E54 2018
621.042--dc23

For more information regarding Larsen and Keller Education and its products, please visit the publisher's website www.larsen-keller.com

Table of Contents

Preface

The process of collecting energy from sources like wind, solar, thermal and other energy resources, in order to store it in electronic devices is called energy harvesting. The different methods of energy harvesting are photovoltaic energy harvesting, wireless energy harvesting, piezoelectric energy harvesting, pyroelectric energy harvesting, etc. This book explores all the important aspects of energy harvesting in the present day scenario. It unfolds the innovative aspects of this field which will be crucial for the holistic understanding of the subject matter. This textbook is meant for students who are looking for an elaborate reference guide on energy harvesting and have the desire to understand the various applications and models of the subject.

A detailed account of the significant topics covered in this book is provided below:

Chapter 1- Energy harvesting is the energy that is derived from thermal, wind, kinetic or solar energy. Some of the devices powered by energy harvesting are wristwatches, thermoelectric generators, micro wind turbine, special antennas, etc. The chapter on energy harvesting offers an insightful focus, keeping in mind the complex subject matter.

Chapter 2- Wind turbines use wind energy to convert it into electric energy. Small wind turbines, unlike commercial wind turbines, are used for microgenerations. Piezoelectricity, thermoelectric generator and rectenna are devices used in energy harvesting. This chapter helps the readers in developing a better understanding related to the devices used in energy harvesting.

Chapter 3- Photovoltaics is the term used for the transformation of light into electricity. The process is accomplished by using semiconducting materials. Concentrator photovoltaics, photovoltaic thermal hybrid solar collector, solar cell, solar cell efficiency and solar panels on spacecraft are some important topics related to the subject of photovoltaics. This chapter will provide an integrated understanding of photovoltaics.

Chapter 4- Energy conversion is the act of changing energy. The change that occurs converts energy from one form to another. This energy can be harvested to ultimately power devices. The aspects elucidated in this chapter are of vital importance, and provide a better understanding of energy conversion.

I would like to make a special mention of my publisher who considered me worthy of this opportunity and also supported me throughout the process. I would also like to thank the editing team at the back-end who extended their help whenever required.

Editor

An Introduction to Energy Harvesting

Energy harvesting is the energy that is derived from thermal, wind, kinetic or solar energy. Some of the devices powered by energy harvesting are wristwatches, thermoelectric generators, micro wind turbine, special antennas, etc. The chapter on energy harvesting offers an insightful focus, keeping in mind the complex subject matter.

Energy Harvesting

Energy harvesting (also known as power harvesting or energy scavenging or ambient power) is the process by which energy is derived from external sources (e.g., solar power, thermal energy, wind energy, salinity gradients, and kinetic energy, also known as ambient energy), captured, and stored for small, wireless autonomous devices, like those used in wearable electronics and wireless sensor networks.

Energy harvesters provide a very small amount of power for low-energy electronics. While the input fuel to some large-scale generation costs resources (oil, coal, etc.), the energy source for energy harvesters is present as ambient background. For example, temperature gradients exist from the operation of a combustion engine and in urban areas, there is a large amount of electromagnetic energy in the environment because of radio and television broadcasting.

One of the earliest applications of ambient power collected from ambient electromagnetic radiation (EMR) is the crystal radio.

The principles of energy harvesting from ambient EMR can be demonstrated with basic components.

Operation

Energy harvesting devices converting ambient energy into electrical energy have attracted much interest in both the military and commercial sectors. Some systems convert motion, such as that of ocean waves, into electricity to be used by oceanographic monitoring sensors for autonomous operation. Future applications may include high power output devices (or arrays of such devices) deployed at remote locations to serve as reliable power stations for large systems. Another application is in wearable electronics, where energy harvesting devices can power or recharge cellphones, mobile computers, radio communication equipment, etc. All of these devices must be sufficiently robust to endure long-term exposure to hostile environments and have a broad range of dynamic sensitivity to exploit the entire spectrum of wave motions.

Accumulating Energy

Energy can also be harvested to power small autonomous sensors such as those developed using

MEMS technology. These systems are often very small and require little power, but their applications are limited by the reliance on battery power. Scavenging energy from ambient vibrations, wind, heat or light could enable smart sensors to be functional indefinitely. Several academic and commercial groups have been involved in the analysis and development of vibration-powered energy harvesting technology, including the Control and Power Group and Optical and Semiconductor Devices Group at Imperial College London, IMEC and the partnering Holst Centr, AdaptivEnergy, LLC, ARVENI, MIT Boston, Victoria University of Wellington, Georgia Tech, UC Berkeley, Southampton University, University of Bristol, Micro Energy System Lab at The University of Tokyo, Nanyang Technological University, PMG Perpetuum, ReVibe Energy, Vestfold University College, National University of Singapore, NiPS Laboratory at the University of Perugia, Columbia University, Universidad Autónoma de Barcelona and USN & Renewable Energy Lab at the University of Ulsan (Ulsan, South Korea). The National Science Foundation also supports an Industry/University Cooperative Research Center led by Virginia Tech and The University of Texas at Dallas called the Center for Energy Harvesting Materials and Systems.

Typical power densities available from energy harvesting devices are highly dependent upon the specific application (affecting the generator's size) and the design itself of the harvesting generator. In general, for motion powered devices, typical values are a few $\mu W/cm^3$ for human body powered applications and hundreds of $\mu W/cm^3$ for generators powered from machinery. Most energy scavenging devices for wearable electronics generate very little power.

Storage of Power

In general, energy can be stored in a capacitor, super capacitor, or battery. Capacitors are used when the application needs to provide huge energy spikes. Batteries leak less energy and are therefore used when the device needs to provide a steady flow of energy.

Use of the Power

Current interest in low power energy harvesting is for independent sensor networks. In these applications an energy harvesting scheme puts power stored into a capacitor then boosted/regulated to a second storage capacitor or battery for the use in the microprocessor. The power is usually used in a sensor application and the data stored or is transmitted possibly through a wireless method.

Motivation

The history of energy harvesting dates back to the windmill and the waterwheel. People have searched for ways to store the energy from heat and vibrations for many decades. One driving force behind the search for new energy harvesting devices is the desire to power sensor networks and mobile devices without batteries. Energy harvesting is also motivated by a desire to address the issue of climate change and global warming.

Devices

There are many small-scale energy sources that generally cannot be scaled up to industrial size:

- Some wristwatches are powered by kinetic energy (called automatic watches), in this case movement of the arm is used. The arm movement causes winding of its mainspring. A newer design introduced by Seiko ("Kinetic") uses movement of a magnet in the electromagnetic generator instead to power the quartz movement. The motion provides a rate of change of flux, which results in some induced emf on the coils. The concept is simply related to Faraday's Law.

- Photovoltaics is a method of generating electrical power by converting solar radiation (both indoors and outdoors) into direct current electricity using semiconductors that exhibit the photovoltaic effect. Photovoltaic power generation employs solar panels composed of a number of cells containing a photovoltaic material. Note that photovoltaics have been scaled up to industrial size and that large solar farms exist.

- Thermoelectric generators (TEGs) consist of the junction of two dissimilar materials and the presence of a thermal gradient. Large voltage outputs are possible by connecting many junctions electrically in series and thermally in parallel. Typical performance is 100-300 μV/K per junction. These can be utilized to capture mW.s of energy from industrial equipment, structures, and even the human body. They are typically coupled with heat sinks to improve temperature gradient.

- Micro wind turbine are used to harvest wind energy readily available in the environment in the form of kinetic energy to power the low power electronic devices such as wireless sensor nodes. When air flows across the blades of the turbine, a net pressure difference is developed between the wind speeds above and below the blades. This will result in a lift force generated which in turn rotate the blades. Similar to photovoltaics, wind farms have been constructed on an industrial scale and are being used to generate substantial amounts of electrical energy.

- Piezoelectric crystals or fibers generate a small voltage whenever they are mechanically deformed. Vibration from engines can stimulate piezoelectric materials, as can the heel of a shoe, or the pushing of a button.

- Special antennas can collect energy from stray radio waves, this can also be done with a Rectenna and theoretically at even higher frequency EM radiation with a Nantenna.

- Power from keys pressed during use of a portable electronic device or remote controller, using magnet and coil or piezoelectric energy converters, may be used to help power the device.

Ambient-radiation Sources

A possible source of energy comes from ubiquitous radio transmitters. Historically, either a large collection area or close proximity to the radiating wireless energy source is needed to get useful power levels from this source. The nantenna is one proposed development which would overcome this limitation by making use of the abundant natural radiation (such as solar radiation).

One idea is to deliberately broadcast RF energy to power remote devices: This is now commonplace in passive radio-frequency identification (RFID) systems, but the Safety and US Federal Communications Commission (and equivalent bodies worldwide) limit the maximum power that can be transmitted this way to civilian use. This method has been used to power individual nodes in a wireless sensor network.

Fluid Flow

Airflow can be harvested by various turbine and non-turbine generator technologies. For example, Zephyr Energy Corporation's patented Windbeam micro generator captures energy from airflow to recharge batteries and power electronic devices. The Windbeam's novel design allows it to operate silently in wind speeds as low as 2 mph. The generator consists of a lightweight beam suspended by durable long-lasting springs within an outer frame. The beam oscillates rapidly when exposed to airflow due to the effects of multiple fluid flow phenomena. A linear alternator assembly converts the oscillating beam motion into usable electrical energy. A lack of bearings and gears eliminates frictional inefficiencies and noise. The generator can operate in low-light environments unsuitable for solar panels (e.g. HVAC ducts) and is inexpensive due to low cost components and simple construction. The scalable technology can be optimized to satisfy the energy requirements and design constraints of a given application.

The flow of blood can also be used to power devices. For instance, the pacemaker developed at the University of Bern, uses blood flow to wind up a spring which in turn drives an electrical micro-generator.

Photovoltaic

Photovoltaic (PV) energy harvesting wireless technology offers significant advantages over wired or solely battery-powered sensor solutions: virtually inexhaustible sources of power with little or no adverse environmental effects. Indoor PV harvesting solutions have to date been powered by specially tuned amorphous silicon (aSi)a technology most used in Solar Calculators. In recent years new PV technologies have come to the forefront in Energy Harvesting such as Dye Sensitized Solar Cells (DSSC). The dyes absorbs light much like chlorophyll does in plants. Electrons released on impact escape to the layer of TiO_2 and from there diffuse, through the electrolyte, as the dye can be tuned to the visible spectrum much higher power can be produced. At 200 lux a DSSC can provide over 10 μW per cm².

picture of a batteryless and wireless wallswitch

Piezoelectric

The piezoelectric effect converts mechanical strain into electric current or voltage. This strain can come from many different sources. Human motion, low-frequency seismic vibrations, and acoustic noise are everyday examples. Except in rare instances the piezoelectric effect operates in AC requiring time-varying inputs at mechanical resonance to be efficient.

Most piezoelectric electricity sources produce power on the order of milliwatts, too small for system application, but enough for hand-held devices such as some commercially available self-winding wristwatches. One proposal is that they are used for micro-scale devices, such as in a device harvesting micro-hydraulic energy. In this device, the flow of pressurized hydraulic fluid drives a reciprocating piston supported by three piezoelectric elements which convert the pressure fluctuations into an alternating current.

As piezo energy harvesting has been investigated only since the late 1990s, it remains an emerging technology. Nevertheless, some interesting improvements were made with the self-powered electronic switch at INSA school of engineering, implemented by the spin-off Arveni. In 2006, the proof of concept of a battery-less wireless doorbell push button was created, and recently, a product showed that classical wireless wallswitch can be powered by a piezo harvester. Other industrial applications appeared between 2000 and 2005, to harvest energy from vibration and supply sensors for example, or to harvest energy from shock.

Piezoelectric systems can convert motion from the human body into electrical power. DARPA has funded efforts to harness energy from leg and arm motion, shoe impacts, and blood pressure for low level power to implantable or wearable sensors. The nanobrushes are another example of a piezoelectric energy harvester. They can be integrated into clothing. Multiple other nanostructures have been exploited to build an energy-harvesting device, for example, a single crystal PMN-PT nanobelt was fabricated and assembled into a piezoelectric energy harvester in 2016. Careful design is needed to minimise user discomfort. These energy harvesting sources by association affect the body. The Vibration Energy Scavenging Project is another project that is set up to try to scavenge electrical energy from environmental vibrations and movements. Microbelt can be used to gather electricity from respiration. Finally, a millimeter-scale piezoelectric energy harvester has also already been created.

The use of piezoelectric materials to harvest power has already become popular. Piezoelectric materials have the ability to transform mechanical strain energy into electrical charge. Piezo elements are being embedded in walkways to recover the "people energy" of footsteps. They can also be embedded in shoes to recover "walking energy". Researchers at MIT developed the first micro-scale piezoelectric energy harvester using thin film PZT in 2005. Arman Hajati and Sang-Gook Kim invented the Ultra Wide-Bandwidth micro-scale piezoelectric energy harvesting device by exploiting the nonlinear stiffness of a doubly clamped microelectromechanical systems (MEMSs) resonator. The stretching strain in a doubly clamped beam shows a nonlinear stiffness, which provides a passive feedback and results in amplitude-stiffened Duffing mode resonance.

Energy from Smart Roads and Piezoelectricity

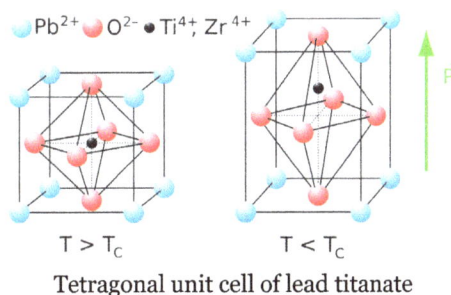

Tetragonal unit cell of lead titanate

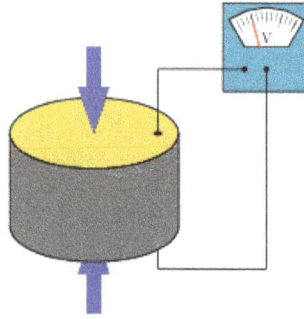

A piezoelectric disk generates a voltage when deformed (change in shape is greatly exaggerated).

Brothers Pierre Curie and Jacques Curie gave the concept of piezoelectric effect in 1880. Piezoelectric effect converts mechanical strain into voltage or electric current and generates electric energy from motion, weight, vibration and temperature changes as shown in the figure.

Considering piezoelectric effect in thin film lead zirconate titanate $Pb(Zr,Ti)O3$ PZT, microelectromechanical systems (MEMS) power generating device has been developed. During recent improvement in piezoelectric technology, Aqsa Abbasi (*also known as Aqsa Aitbar, General secretory at IMS, IEEE MUET Chapter and Director Media at HYD MUN*) diffentiated two modes called d_{31} and d_{33} in vibration converters and re-designed to resonate at specific frequencies from an external vibration energy source, thereby creating electrical energy via the piezoelectric effect using electromechanical damped mass. However, Aqsa further developed beam-structured electrostatic devices that are more difficult to fabricate than PZT MEMS devices versus a similar because general silicon processing involves many more mask steps that do not require PZT film. Piezoelectric d_{31} type sensors and actuators have a cantilever beam structure that consists of a membrane bottom electrode, film, piezoelectric film, and top electrode. More than (*3~5 masks*) mask steps are required for patterning of each layer while have very low induced voltage. Pyroelectric crystals that have a unique polar axis and have spontaneous polarization, along which the spontaneous polarization exists. These are the crystals of classes *6mm, 4mm, mm2, 6, 4, 3m, 3,2, m*. The special polar axis—crystallophysical axis $X3$ — coincides with the axes *L6,L4, L3,* and *L2* of the crystals or lies in the unique straight plane *P (class "m")*. Consequently, the electric centers of positive and negative charges are displaced of an elementary cell from equilibrium positions, i.e., the spontaneous polarization of the crystal changes. Therefore, all considered crystals have spontaneous polarization $Ps = P3$. Since piezoelectric effect in pyroelectric crystals arises as a result of changes in their spontaneous polarization under external effects (electric fields, mechanical stresses). As a result of displacement, Aqsa Abbasi introduced change in the components ΔP_s along all three axes $\Delta P_s = (\Delta P_1, \Delta P_2, \Delta P_3)$. Suppose that $\Delta P_s = (\Delta P_1, \Delta P_2, \Delta P_3)$ is proportional to the mechanical stresses causing in a first approximation, which results $\Delta P_i = diklTkl$ where *Tkl* represents the mechanical stress and *dikl* represents the piezoelectric modules.

PZT thin films have attracted attention for applications such as force sensors, accelerometers, gyroscopes actuators, tunable optics, micro pumps, ferroelectric RAM, display systems and smart roads, when energy sources are limited, energy harvesting plays an important role in the environment. Smart roads have the potential to play an important role in power generation. Embedding piezoelectric material in the road can convert pressure exerted by moving vehicles into voltage and current.

Smart Transportation Intelligent System

Piezoelectric sensors are most useful in Smart-road technologies that can be used to create systems that are intelligent and improve productivity in the long run. Imagine highways that alert motorists of a traffic jam before it forms. Or bridges that report when they are at risk of collapse, or an electric grid that fixes itself when blackouts hit. For many decades, scientists and experts have argued that the best way to fight congestion is intelligent transportation systems, such as roadside sensors to measure traffic and synchronized traffic lights to control the flow of vehicles. But the spread of these technologies has been limited by cost. There are also some other smart-technology shovel ready projects which could be deployed fairly quickly, but most of the technologies are still at the development stage and might not be practically available for five years or more.

Pyroelectric

The pyroelectric effect converts a temperature change into electric current or voltage. It is analogous to the piezoelectric effect, which is another type of ferroelectric behavior. Pyroelectricity requires time-varying inputs and suffers from small power outputs in energy harvesting applications due to its low operating frequencies. However, one key advantage of pyroelectrics over thermoelectrics is that many pyroelectric materials are stable up to 1200 ^{0}C or higher, enabling energy harvesting from high temperature sources and thus increasing thermodynamic efficiency.

One way to directly convert waste heat into electricity is by executing the Olsen cycle on pyroelectric materials. The Olsen cycle consists of two isothermal and two isoelectric field processes in the electric displacement-electric field (D-E) diagram. The principle of the Olsen cycle is to charge a capacitor via cooling under low electric field and to discharge it under heating at higher electric field. Several pyroelectric converters have been developed to implement the Olsen cycle using conduction, convection, or radiation. It has also been established theoretically that pyroelectric conversion based on heat regeneration using an oscillating working fluid and the Olsen cycle can reach Carnot efficiency between a hot and a cold thermal reservoir. Moreover, recent studies have established polyvinylidene fluoride trifluoroethylene [P(VDF-TrFE)] polymers and lead lanthanum zirconate titanate (PLZT) ceramics as promising pyroelectric materials to use in energy converters due to their large energy densities generated at low temperatures. Additionally, a pyroelectric scavenging device that does not require time-varying inputs was recently introduced. The energy-harvesting device uses the edge-depolarizing electric field of a heated pyroelectric to convert heat energy into mechanical energy instead of drawing electric current off two plates attached to the crystal-faces.

Thermoelectrics

In 1821, Thomas Johann Seebeck discovered that a thermal gradient formed between two dissimilar conductors produces a voltage. At the heart of the thermoelectric effect is the fact that a temperature gradient in a conducting material results in heat flow; this results in the diffusion of charge carriers. The flow of charge carriers between the hot and cold regions in turn creates a voltage difference. In 1834, Jean Charles Athanase Peltier discovered that running an electric current through the junction of two dissimilar conductors could, depending on the direction of the current, cause it to act as a heater or cooler. The heat absorbed or produced is proportional to the current, and the proportionality constant is known as the Peltier coefficient. Today, due to knowledge of the Seebeck and Peltier effects, thermoelectric materials can be used as heaters, coolers and generators (TEGs).

Ideal thermoelectric materials have a high Seebeck coefficient, high electrical conductivity, and low thermal conductivity. Low thermal conductivity is necessary to maintain a high thermal gradient at the junction. Standard thermoelectric modules manufactured today consist of P- and N-doped bismuth-telluride semiconductors sandwiched between two metallized ceramic plates. The ceramic plates add rigidity and electrical insulation to the system. The semiconductors are connected electrically in series and thermally in parallel.

Miniature thermocouples have been developed that convert body heat into electricity and generate 40μW at 3V with a 5 degree temperature gradient, while on the other end of the scale, large thermocouples are used in nuclear RTG batteries.

Practical examples are the finger-heartratemeter by the Holst Centre and the thermogenerators by the Fraunhofer Gesellschaft.

Advantages to thermoelectrics:

1. No moving parts allow continuous operation for many years. *Tellurex Corporation* (a thermoelectric production company) claims that thermoelectrics are capable of over 100,000 hours of steady state operation.

2. Thermoelectrics contain no materials that must be replenished.

3. Heating and cooling can be reversed.

One downside to thermoelectric energy conversion is low efficiency (currently less than 10%). The development of materials that are able to operate in higher temperature gradients, and that can conduct electricity well without also conducting heat (something that was until recently thought impossible), will result in increased efficiency.

Future work in thermoelectrics could be to convert wasted heat, such as in automobile engine combustion, into electricity.

Electrostatic (Capacitive)

This type of harvesting is based on the changing capacitance of vibration-dependent capacitors. Vibrations separate the plates of a charged variable capacitor, and mechanical energy is converted into electrical energy. Electrostatic energy harvesters need a polarization source to work and to convert mechanical energy from vibrations into electricity. The polarization source should be in the order of some hundreds of volts; this greatly complicates the power management circuit. Another solution consists in using electrets, that are electrically charged dielectrics able to keep the polarization on the capacitor for years. It's possible to adapt structures from classical electrostatic induction generators, which also extract energy from variable capacitances, for this purpose. The resulting devices are self-biasing, and can directly charge batteries, or can produce exponentially growing voltages on storage capacitors, from which energy can be periodically extracted by DC/ DC converters.

Magnetic Induction

Magnets wobbling on a cantilever are sensitive to even small vibrations and generate microcurrents by moving relative to conductors due to Faraday's law of induction. By developing a miniature device of this kind in 2007, a team from the University of Southampton made possible the planting of such a device in environments that preclude having any electrical connection to the outside world. Sensors in inaccessible places can now generate their own power and transmit data to outside receivers.

One of the major limitations of the magnetic vibration energy harvester developed at University of Southampton is the size of the generator, in this case approximately one cubic centimeter, which is much too large to integrate into today's mobile technologies. The complete generator including circuitry is a massive 4 cm by 4 cm by 1 cm nearly the same size as some mobile devices such as the iPod nano. Further reductions in the dimensions are possible through the integration of new and more flexible materials as the cantilever beam component. In 2012, a group at Northwestern University developed a vibration-powered generator out of polymer in the form of a spring. This device was able to target the same frequencies as the University of Southampton groups silicon based device but with one third the size of the beam component.

A new approach to magnetic induction based energy harvesting has also been proposed by using ferrofluids. The journal article, *Electromagnetic ferrofluid-based energy harvester*, discusses the use of ferrofluids to harvest low frequency vibrational energy at 2.2Hz with a power output of ~80mW per g.

Commercially successful vibration energy harvesters have been developed from the early University of Southampton prototypes by Perpetuum. These have to be sufficiently large to generate the power required by wireless sensor nodes (wsn)but in M2M applications this is not normally an issue. These harvesters are now being supplied in large volumes to power wsn's made by companies such as GE and Emerson and also for train bearing monitoring systems made by Perpetuum. Overhead powerline sensors can use magnetic induction to harvest energy directly from the conductor they are monitoring.

Blood Sugar

Another way of energy harvesting is through the oxidation of blood sugars. These energy harvesters are called biobatteries. They could be used to power implanted electronic devices (e.g., pacemakers, implanted biosensors for diabetics, implanted active RFID devices, etc.). At present, the Minteer Group of Saint Louis University has created enzymes that could be used to generate power from blood sugars. However, the enzymes would still need to be replaced after a few years. In 2012, a pacemaker was powered by implantable biofuel cells at Clarkson University under the leadership of Dr. Evgeny Katz.

Tree-based

Tree metabolic energy harvesting is a type of bio-energy harvesting. Voltree has developed a method for harvesting energy from trees. These energy harvesters are being used to power remote sensors and mesh networks as the basis for a long term deployment system to monitor forest fires and weather in the forest. Their website says that the useful life of such a device should be limited only by the lifetime of the tree to which it is attached. They recently deployed a small test network in a US National Park forest.

Other sources of energy from trees include capturing the physical movement of the tree in a generator. Theoretical analysis of this source of energy shows some promise in powering small electronic devices. A practical device based on this theory has been built and successfully powered a sensor node for a year.

Metamaterial

A metamaterial-based device wirelessly converts a 900 MHz microwave signal to 7.3 volts of direct current (greater than that of a USB device). The device can be tuned to harvest other signals including Wi-Fi signals, satellite signals, or even sound signals. The experimental device used a series of five fiberglass and copper conductors. Conversion efficiency reached 37 percent. When traditional antennas are close to each other in space they interfere with each other. But since RF power goes down by the cube of the distance, the amount of power is very very small. While the claim of 7.3 volts is grand, the measurement is for an open circuit. Since the power is so low, there can be almost no current when any load is attached.

Atmospheric Pressure Changes

The change in air pressure due to temperature changes or weather patterns vs. a sealed chamber has been used to provide power for mechanical clocks such as the Atmos clock.

Human Power

An athlete can produce about 300 to 400 watts of mechanical power for an hour or so (1/3 kWh/1/2 hp), but adults of good average fitness average between 50 and 150 watts for an hour of vigorous exercise (1/10 kWh). A healthy laborer may sustain an average output of about 75 watts for some eight hours (½ Kwh). Pedal power is therefore most suitable for fairly short tasks with modest power demand.

Body Accessories

Biomechanical energy harvesters are also being created. One current model is the biomechanical energy harvester of Max Donelan which straps around the knee. Devices as this allow the generation of 2.5 watts of power per knee. This is enough to power some 5 cell phones. The Soccket can generate and store 6 watts. There is also a knee brace developed by Bionic Power which is based in Canada.

Body-energy can also be extracted as described for wristwatches, from blood for pacemakers.

Pavement

A company called PaveGen produces pavement slabs that produce electricity; in addition to permanent installations it has demonstrated at various events such as the 2012 London Olympics and the Paris Marathon.

Pedal Power

Pedal power is simple, efficient, and practical. There are essentially two designs, the reciprocating treadle and the rotating pedal crankset.

Stationary machinery such as the bodger pole lathe have been in use for several thousands of years (since at least the bronze age) and exactly the same reciprocating treadle mechanism, with somewhat more advanced mechanics, was adapted to sewing machines as patented by Isaac Singer in 1851.

Pedal Power is the most familiar as used for bicycles or tricycles, popular for light transport since the late 19th Century. The beach-side quadracycle which was patented in 1853 demonstrated that power could be drawn from more than one cyclist.

Pedal Electricity Generators

Some stationary bicycles have been fitted with generators and batteries, and at least one US patents has been granted. Usually the amount of useful electrical energy generated or collected is low because neither generators nor batteries are very efficient, and initially power is lost in converting reciprocating muscular power to rotatory force. These problems are surmountable, there are designs to produce up to 120W electrical output for extended time.

Dynapod Mechanical Power

A better solution was proposed as long ago as 1980 by Volunteers in Technical Assistance (VITA Maryland, USA) who called their device a 'dynapod'. Their idea is to re-engineer common Home appliance with small (fractional – less than 1-horsepower) electric motors (c. 500W – 1000W) which are used for short periods, for example food mixers, grinding machines, hand-held power-drills and light wood-working equipment.

Since most domestic appliances are used in relatively static environments, and control of tool-velocity is often important, pedal mechanisms can deliver both muscle power and fine speed control to the where it is needed, while also providing comfortable seating for the user and, additionally, leaving both hands free to manipulate the work-piece or appliance.

Future Directions

Electroactive polymers (EAPs) have been proposed for harvesting energy. These polymers have a large strain, elastic energy density, and high energy conversion efficiency. The total weight of systems based on EAPs(electroactive polymers) is proposed to be significantly lower than those based on piezoelectric materials.

Nanogenerators, such as the one made by Georgia Tech, could provide a new way for powering devices without batteries. As of 2008, it only generates some dozen nanowatts, which is too low for any practical application.

Noise has been the subject of a proposal by NiPS Laboratory in Italy to harvest wide spectrum low scale vibrations via a nonlinear dynamical mechanism that can improve harvester efficiency up to a factor 4 compared to traditional linear harvesters.

Combinations of different types of energy harvesters can further reduce dependence on batteries, particularly in environments where the available ambient energy types change periodically. This type of complementary balanced energy harvesting has the potential to increase reliability of wireless sensor systems for structural health monitoring.

Renewable Energy

Wind, solar, and hydroelectricity are three emerging renewable sources of energy

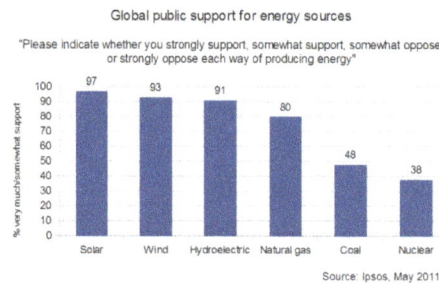

Global public support for different energy sources (2011) based on a poll by Ipsos Global @dvisor

Renewable energy is energy that is collected from renewable resources, which are naturally replenished on a human timescale, such as sunlight, wind, rain, tides, waves, and geothermal heat. Renewable energy often provides energy in four important areas: electricity generation, air and water heating/cooling, transportation, and rural (off-grid) energy services.

Based on REN21's 2016 report, renewables contributed 19.2% to humans' global energy consumption and 23.7% to their generation of electricity in 2014 and 2015, respectively. This energy consumption is divided as 8.9% coming from traditional biomass, 4.2% as heat energy (modern biomass, geothermal and solar heat), 3.9% hydro electricity and 2.2% is electricity from wind, solar, geothermal, and biomass. Worldwide investments in renewable technologies amounted to more than US$286 billion in 2015, with countries like China and the United States heavily investing in wind, hydro, solar and biofuels. Globally, there are an estimated 7.7 million jobs associated with the renewable energy industries, with solar photovoltaics being the largest renewable employer. As of 2015 worldwide, more than half of all new electricity capacity installed was renewable.

Renewable energy resources exist over wide geographical areas, in contrast to other energy sources, which are concentrated in a limited number of countries. Rapid deployment of renewable energy and energy efficiency is resulting in significant energy security, climate change mitigation, and economic benefits. The results of a recent review of the literature concluded that as greenhouse gas (GHG) emitters begin to be held liable for damages resulting from GHG emissions resulting in climate change, a high value for liability mitigation would provide powerful incentives for deployment of renewable energy technologies. In international public opinion surveys there is strong support for promoting renewable sources such as solar power and wind power. At the national level, at least 30 nations around the world already have renewable energy contributing more than 20 percent of energy supply. National renewable energy markets are projected to continue to grow strongly in the coming decade and beyond. Some places and at least two countries, Iceland and Norway generate all their electricity using renewable energy already, and many other countries have the set a goal to reach 100% renewable energy in the future. For example, in Denmark the government decided to switch the total energy supply (electricity, mobility and heating/cooling) to 100% renewable energy by 2050.

While many renewable energy projects are large-scale, renewable technologies are also suited to rural and remote areas and developing countries, where energy is often crucial in human development. United Nations' Secretary-General Ban Ki-moon has said that renewable energy has the ability to lift the poorest nations to new levels of prosperity. As most of renewables provide electricity, renewable energy deployment is often applied in conjunction with further electrification, which has several benefits: Electricity can be converted to heat (where necessary generating higher temperatures than fossil fuels), can be converted into mechanical energy with high efficiency and is clean at the point of consumption. In addition to that electrification with renewable energy is much more efficient and therefore leads to a significant reduction in primary energy requirements, because most renewables don't have a steam cycle with high losses (fossil power plants usually have losses of 40 to 65%).

Renewable energy systems are rapidly becoming more efficient and cheaper. Their share of total energy consumption is increasing. Growth in consumption of coal and oil could end by 2020 due to increased uptake of renewables and natural gas.

Overview

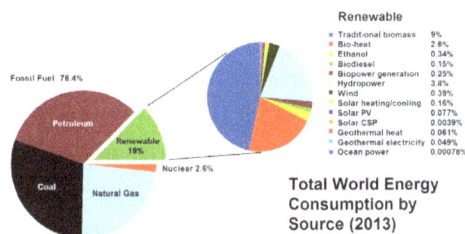

World energy consumption by source. Renewables accounted for 19% in 2012.

PlanetSolar, the world's largest solar-powered boat and the first ever solar electric vehicle to circumnavigate the globe (in 2012)

Renewable energy flows involve natural phenomena such as sunlight, wind, tides, plant growth, and geothermal heat, as the International Energy Agency explains:

Renewable energy is derived from natural processes that are replenished constantly. In its various forms, it derives directly from the sun, or from heat generated deep within the earth. Included in the definition is electricity and heat generated from solar, wind, ocean, hydropower, biomass, geothermal resources, and biofuels and hydrogen derived from renewable resources.

Renewable energy resources and significant opportunities for energy efficiency exist over wide geographical areas, in contrast to other energy sources, which are concentrated in a limited number of countries. Rapid deployment of renewable energy and energy efficiency, and technological diversification of energy sources, would result in significant energy security and economic benefits. It would also reduce environmental pollution such as air pollution caused by burning of fossil fuels and improve public health, reduce premature mortalities due to pollution and save associated health costs that amount to several hundred billion dollars annually only in the United States. Renewable energy sources, that derive their energy from the sun, either directly or indirectly, such as hydro and wind, are expected to be capable of supplying humanity energy for almost another 1 billion years, at which point the predicted increase in heat from the sun is expected to make the surface of the earth too hot for liquid water to exist.

Climate change and global warming concerns, coupled with high oil prices, peak oil, and increasing government support, are driving increasing renewable energy legislation, incentives and commercialization. New government spending, regulation and policies helped the industry weather the global financial crisis better than many other sectors. According to a 2011 projection by the International Energy Agency, solar power generators may produce most of the world's electricity within 50 years, reducing the emissions of greenhouse gases that harm the environment.

As of 2011, small solar PV systems provide electricity to a few million households, and micro-hydro configured into mini-grids serves many more. Over 44 million households use biogas made in household-scale digesters for lighting and/or cooking, and more than 166 million households rely on a new generation of more-efficient biomass cookstoves. United Nations' Secretary-General Ban Ki-moon has said that renewable energy has the ability to lift the poorest nations to new levels of prosperity. At the national level, at least 30 nations around the world already have renewable energy contributing more than 20% of energy supply. National renewable energy markets are projected to continue to grow strongly in the coming decade and beyond, and some 120 countries have various policy targets for longer-term shares of renewable energy, including a 20% target of all electricity generated for the European Union by 2020. Some countries have much higher long-term policy targets of up to 100% renewables. Outside Europe, a diverse group of 20 or more other countries target renewable energy shares in the 2020–2030 time frame that range from 10% to 50%.

Renewable energy often displaces conventional fuels in four areas: electricity generation, hot water/space heating, transportation, and rural (off-grid) energy services:

- Power generation: By 2040, renewable energy is projected to equal coal and natural gas electricity generation. Several jurisdictions, including Denmark, Germany, the state of South Australia and some US states have achieved high integration of variable renewables. For example, in 2015 wind power met 42% of electricity demand in Denmark, 23.2% in

Portugal and 15.5% in Uruguay. Interconnectors enable countries to balance electricity systems by allowing the import and export of renewable energy. Innovative hybrid systems have emerged between countries and regions.

- Heating: Solar water heating makes an important contribution to renewable heat in many countries, most notably in China, which now has 70% of the global total (180 GWth). Most of these systems are installed on multi-family apartment buildings and meet a portion of the hot water needs of an estimated 50–60 million households in China. Worldwide, total installed solar water heating systems meet a portion of the water heating needs of over 70 million households. The use of biomass for heating continues to grow as well. In Sweden, national use of biomass energy has surpassed that of oil. Direct geothermal for heating is also growing rapidly. The newest addition to Heating is from Geothermal Heat Pumps which provide both heating and cooling, and also flatten the electric demand curve and are thus an increasing national priority.

- Transportation: Bioethanol is an alcohol made by fermentation, mostly from carbohydrates produced in sugar or starch crops such as corn, sugarcane, or sweet sorghum. Cellulosic biomass, derived from non-food sources such as trees and grasses is also being developed as a feedstock for ethanol production. Ethanol can be used as a fuel for vehicles in its pure form, but it is usually used as a gasoline additive to increase octane and improve vehicle emissions. Bioethanol is widely used in the USA and in Brazil. Biodiesel can be used as a fuel for vehicles in its pure form, but it is usually used as a diesel additive to reduce levels of particulates, carbon monoxide, and hydrocarbons from diesel-powered vehicles. Biodiesel is produced from oils or fats using transesterification and is the most common biofuel in Europe.

A bus fueled by biodiesel

A solar vehicle is an electric vehicle powered completely or significantly by direct solar energy. Usually, photovoltaic (PV) cells contained in solar panels convert the sun's energy directly into electric energy. The term "solar vehicle" usually implies that solar energy is used to power all or part of a vehicle's propulsion. Solar power may be also used to provide power for communications or controls or other auxiliary functions. Solar vehicles are not sold as practical day-to-day transportation devices at present, but are primarily demonstration vehicles and engineering exercises, often sponsored by government agencies. However, indirectly solar-charged vehicles are widespread and solar boats are available commercially.

History

Prior to the development of coal in the mid 19th century, nearly all energy used was renewable. Almost without a doubt the oldest known use of renewable energy, in the form of traditional biomass to fuel fires, dates from 790,000 years ago. Use of biomass for fire did not become commonplace until many hundreds of thousands of years later, sometime between 200,000 and 400,000 years ago. Probably the second oldest usage of renewable energy is harnessing the wind in order to drive ships over water. This practice can be traced back some 7000 years, to ships on the Nile. Moving into the time of recorded history, the primary sources of traditional renewable energy were human labor, animal power, water power, wind, in grain crushing windmills, and firewood, a traditional biomass. A graph of energy use in the United States up until 1900 shows oil and natural gas with about the same importance in 1900 as wind and solar played in 2010.

In the 1860s and '70s there were already fears that civilization would run out of fossil fuels and the need was felt for a better source. In 1873 Professor Augustine Mouchot wrote:

The time will arrive when the industry of Europe will cease to find those natural resources, so necessary for it. Petroleum springs and coal mines are not inexhaustible but are rapidly diminishing in many places. Will man, then, return to the power of water and wind? Or will he emigrate where the most powerful source of heat sends its rays to all? History will show what will come.

In 1885, Werner von Siemens, commenting on the discovery of the photovoltaic effect in the solid state, wrote:

In conclusion, I would say that however great the scientific importance of this discovery may be, its practical value will be no less obvious when we reflect that the supply of solar energy is both without limit and without cost, and that it will continue to pour down upon us for countless ages after all the coal deposits of the earth have been exhausted and forgotten.

Max Weber mentioned the end of fossil fuel in the concluding paragraphs of his Die protestantische Ethik und der Geist des Kapitalismus, published in 1905.

Development of solar engines continued until the outbreak of World War I. The importance of solar energy was recognized in a 1911 *Scientific American* article: "in the far distant future, natural fuels having been exhausted [solar power] will remain as the only means of existence of the human race".

The theory of peak oil was published in 1956. In the 1970s environmentalists promoted the development of renewable energy both as a replacement for the eventual depletion of oil, as well as for an escape from dependence on oil, and the first electricity generating wind turbines appeared. Solar had long been used for heating and cooling, but solar panels were too costly to build solar farms until 1980.

The IEA 2014 World Energy Outlook projects a growth of renewable energy supply from 1,700 gigawatts in 2014 to 4,550 gigawatts in 2040. Fossil fuels received about $550 billion in subsidies in 2013, compared to $120 billion for all renewable energies.

Mainstream Technologies

Wind Power

The 845 MW Shepherds Flat Wind Farm near Arlington, Oregon, USA.

Airflows can be used to run wind turbines. Modern utility-scale wind turbines range from around 600 kW to 5 MW of rated power, although turbines with rated output of 1.5–3 MW have become the most common for commercial use. The largest generator capacity of a single installed onshore wind turbine reached 7.5 MW in 2015. The power available from the wind is a function of the cube of the wind speed, so as wind speed increases, power output increases up to the maximum output for the particular turbine. Areas where winds are stronger and more constant, such as offshore and high altitude sites, are preferred locations for wind farms. Typically full load hours of wind turbines vary between 16 and 57 percent annually, but might be higher in particularly favorable offshore sites.

Wind-generated electricity met nearly 4% of global electricity demand in 2015, with nearly 63 GW of new wind power capacity installed. Wind energy was the leading source of new capacity in Europe, the US and Canada, and the second largest in China. In Denmark, wind energy met more than 40% of its electricity demand while Ireland, Portugal and Spain each met nearly 20%.

Globally, the long-term technical potential of wind energy is believed to be five times total current global energy production, or 40 times current electricity demand, assuming all practical barriers needed were overcome. This would require wind turbines to be installed over large areas, particularly in areas of higher wind resources, such as offshore. As offshore wind speeds average ~90% greater than that of land, so offshore resources can contribute substantially more energy than land stationed turbines. In 2014 global wind generation was 706 terawatt-hours or 3% of the worlds total electricity.

Hydropower

The Three Gorges Dam on the Yangtze River in China.

In 2015 hydropower generated 16.6% of the worlds total electricity and 70% of all renewable electricity. Since water is about 800 times denser than air, even a slow flowing stream of water, or moderate sea swell, can yield considerable amounts of energy. There are many forms of water energy:

- Historically hydroelectric power came from constructing large hydroelectric dams and reservoirs, which are still popular in third world countries. The largest of which is the Three Gorges Dam(2003) in China and the Itaipu Dam(1984) built by Brazil and Paraguay.

- Small hydro systems are hydroelectric power installations that typically produce up to 50 MW of power. They are often used on small rivers or as a low impact development on larger rivers. China is the largest producer of hydroelectricity in the world and has more than 45,000 small hydro installations.

- Run-of-the-river hydroelectricity plants derive kinetic energy from rivers without the creation of a large reservoir. This style of generation may still produce a large amount of electricity, such as the Chief Joseph Dam on the Columbia river in the United States.

Hydropower is produced in 150 countries, with the Asia-Pacific region generating 32 percent of global hydropower in 2010. For countries having the largest percentage of electricity from renewables, the top 50 are primarily hydroelectric. China is the largest hydroelectricity producer, with 721 terawatt-hours of production in 2010, representing around 17 percent of domestic electricity use. There are now three hydroelectricity stations larger than 10 GW: the Three Gorges Dam in China, Itaipu Dam across the Brazil/Paraguay border, and Guri Dam in Venezuela.

Wave power, which captures the energy of ocean surface waves, and tidal power, converting the energy of tides, are two forms of hydropower with future potential; however, they are not yet widely employed commercially. A demonstration project operated by the Ocean Renewable Power Company on the coast of Maine, and connected to the grid, harnesses tidal power from the Bay of Fundy, location of world's highest tidal flow. Ocean thermal energy conversion, which uses the temperature difference between cooler deep and warmer surface waters, has currently no economic feasibility.

Solar Energy

Satellite image of the 550-megawatt Topaz Solar Farm in California, USA.

Solar energy, radiant light and heat from the sun, is harnessed using a range of ever-evolving technologies such as solar heating, photovoltaics, concentrated solar power (CSP), concentrator photovoltaics (CPV), solar architecture and artificial photosynthesis. Solar technologies are broadly characterized as either passive solar or active solar depending on the way they capture, convert and distribute solar energy. Passive solar techniques include orienting a building to the Sun, selecting materials with favorable thermal mass or light dispersing properties, and designing spaces that naturally circulate air. Active solar technologies encompass solar thermal energy, using solar collectors for heating, and solar power, converting sunlight into electricity either directly using photovoltaics (PV), or indirectly using concentrated solar power (CSP).

A photovoltaic system converts light into electrical direct current (DC) by taking advantage of the photoelectric effect. Solar PV has turned into a multi-billion, fast-growing industry, continues to improve its cost-effectiveness, and has the most potential of any renewable technologies together with CSP. Concentrated solar power (CSP) systems use lenses or mirrors and tracking systems to focus a large area of sunlight into a small beam. Commercial concentrated solar power plants were first developed in the 1980s. CSP-Stirling has by far the highest efficiency among all solar energy technologies.

In 2011, the International Energy Agency said that "the development of affordable, inexhaustible and clean solar energy technologies will have huge longer-term benefits. It will increase countries' energy security through reliance on an indigenous, inexhaustible and mostly import-independent resource, enhance sustainability, reduce pollution, lower the costs of mitigating climate change, and keep fossil fuel prices lower than otherwise. These advantages are global. Hence the additional costs of the incentives for early deployment should be considered learning investments; they must be wisely spent and need to be widely shared". In 2014 global solar generation was 186 terawatt-hours, slightly less than 1% of the worlds total grid electricity. Italy has the largest proportion of solar electricity in the world, in 2015 solar supplied 7.8% of electricity demand in Italy.

Geothermal Energy

Steam rising from the Nesjavellir Geothermal Power Station in Iceland.

High Temperature Geothermal energy is from thermal energy generated and stored in the Earth. Thermal energy is the energy that determines the temperature of matter. Earth's geothermal energy originates from the original formation of the planet and from radioactive decay of minerals (in currently uncertain but possibly roughly equal proportions). The geothermal gradient, which is the difference in temperature between the core of the planet and its surface, drives a continuous conduction of thermal energy in the form of heat from the core to the surface. The adjective *geothermal* originates from the Greek roots *geo*, meaning earth, and *thermos*, meaning heat.

The heat that is used for geothermal energy can be from deep within the Earth, all the way down to Earth's core – 4,000 miles (6,400 km) down. At the core, temperatures may reach over 9,000 °F (5,000 °C). Heat conducts from the core to surrounding rock. Extremely high temperature and pressure cause some rock to melt, which is commonly known as magma. Magma convects upward since it is lighter than the solid rock. This magma then heats rock and water in the crust, sometimes up to 700 °F (371 °C).

From hot springs, geothermal energy has been used for bathing since Paleolithic times and for space heating since ancient Roman times, but it is now better known for electricity generation.

Low Temperature Geothermal refers to the use of the outer crust of the earth as a Thermal Battery to facilitate Renewable thermal energy for heating and cooling buildings, and other refrigeration and industrial uses. In this form of Geothermal, a Geothermal Heat Pump and Ground-coupled heat exchanger are used together to move heat energy into the earth (for cooling) and out of the earth (for heating) on a varying seasonal basis. Low temperature Geothermal (generally referred to as "GHP") is an increasingly important renewable technology because it both reduces total annual energy loads associated with heating and cooling, and it also flattens the electric demand curve eliminating the extreme summer and winter peak electric supply requirements. Thus Low Temperature Geothermal/GHP is becoming an increasing national priority with multiple tax credit support and focus as part of the ongoing movement toward Net Zero Energy. New York City has even just passed a law to require GHP anytime is shown to be economical with 20 year financing including the Socialized Cost of Carbon.

Bio Energy

Sugarcane plantation to produce ethanol in Brazil

A CHP power station using wood to supply 30,000 households in France.

Biomass is biological material derived from living, or recently living organisms. It most often refers to plants or plant-derived materials which are specifically called lignocellulosic biomass. As an energy source, biomass can either be used directly via combustion to produce heat, or indirectly after converting it to various forms of biofuel. Conversion of biomass to biofuel can be achieved by different methods which are broadly classified into: *thermal*, *chemical*, and *biochemical* methods. Wood remains the largest biomass energy source today; examples include forest residues – such as dead trees, branches and tree stumps –, yard clippings, wood chips and even municipal solid waste. In the second sense, biomass includes plant or animal matter that can be converted into fibers or other industrial chemicals, including biofuels. Industrial biomass can be grown from numerous types of plants, including miscanthus, switchgrass, hemp, corn, poplar, willow, sorghum, sugarcane, bamboo, and a variety of tree species, ranging from eucalyptus to oil palm (palm oil).

Plant energy is produced by crops specifically grown for use as fuel that offer high biomass output per hectare with low input energy. Some examples of these plants are wheat, which typically yield 7.5–8 tonnes of grain per hectare, and straw, which typically yield 3.5–5 tonnes per hectare in the UK. The grain can be used for liquid transportation fuels while the straw can be burned to produce heat or electricity. Plant biomass can also be degraded from cellulose to glucose through a series of chemical treatments, and the resulting sugar can then be used as a first generation biofuel.

Biomass can be converted to other usable forms of energy like methane gas or transportation fuels like ethanol and biodiesel. Rotting garbage, and agricultural and human waste, all release methane gas – also called landfill gas or biogas. Crops, such as corn and sugarcane, can be fermented to produce the transportation fuel, ethanol. Biodiesel, another transportation fuel, can be produced from left-over food products like vegetable oils and animal fats. Also, biomass to liquids (BTLs) and cellulosic ethanol are still under research. There is a great deal of research involving algal fuel or algae-derived biomass due to the fact that it's a non-food resource and can be produced at rates 5 to 10 times those of other types of land-based agriculture, such as corn and soy. Once harvested, it can be fermented to produce biofuels such as ethanol, butanol, and methane, as well as biodiesel and hydrogen. The biomass used for electricity generation varies by region. Forest by-products, such as wood residues, are common in the United States. Agricultural waste is common in Mauritius (sugar cane residue) and Southeast Asia (rice husks). Animal husbandry residues, such as poultry litter, are common in the United Kingdom.

Biofuels include a wide range of fuels which are derived from biomass. The term covers solid, liquid, and gaseous fuels. Liquid biofuels include bioalcohols, such as bioethanol, and oils, such as biodiesel. Gaseous biofuels include biogas, landfill gas and synthetic gas. Bioethanol is an alcohol made by fermenting the sugar components of plant materials and it is made mostly from sugar and starch crops. These include maize, sugarcane and, more recently, sweet sorghum. The latter crop is particularly suitable for growing in dryland conditions, and is being investigated by International Crops Research Institute for the Semi-Arid Tropics for its potential to provide fuel, along with food and animal feed, in arid parts of Asia and Africa.

With advanced technology being developed, cellulosic biomass, such as trees and grasses, are also used as feedstocks for ethanol production. Ethanol can be used as a fuel for vehicles in its pure form, but it is usually used as a gasoline additive to increase octane and improve vehicle emissions. Bioethanol is widely used in the United States and in Brazil. The energy costs for producing bio-ethanol are almost equal to, the energy yields from bio-ethanol. However, according to the European Environment Agency, biofuels do not address global warming concerns. Biodiesel is made from vegetable oils, animal fats or recycled greases. It can be used as a fuel for vehicles in its pure form, or more commonly as a diesel additive to reduce levels of particulates, carbon monoxide, and hydrocarbons from diesel-powered vehicles. Biodiesel is produced from oils or fats using transesterification and is the most common biofuel in Europe. Biofuels provided 2.7% of the world's transport fuel in 2010.

Biomass, biogas and biofuels are burned to produce heat/power and in doing so harm the environment. Pollutants such as sulphurous oxides (SO_x), nitrous oxides (NO_x), and particulate matter (PM) are produced from the combustion of biomass; the World Health Organisation estimates that 7 million premature deaths are caused each year by air pollution. Biomass combustion is a major contributor. The life cycle of the plants is sustainable, the lives of people less so.

Energy Storage

Energy storage is a collection of methods used to store electrical energy on an electrical power grid, or off it. Electrical energy is stored during times when production (especially from intermittent power plants such as renewable electricity sources such as wind power, tidal power, solar power) exceeds consumption, and returned to the grid when production falls below consumption. Pumped-storage hydroelectricity is used for more than 90% of all grid power storage.

Market and Industry Trends

Renewable power has been more effective in creating jobs than coal or oil in the United States.

Growth of Renewables

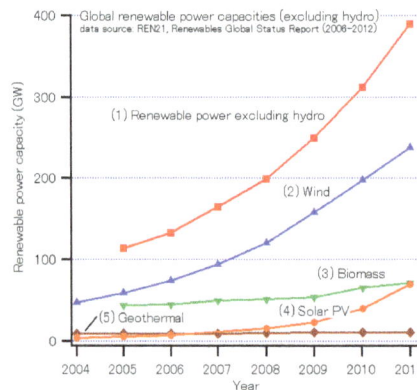

Global growth of renewables through to 2011

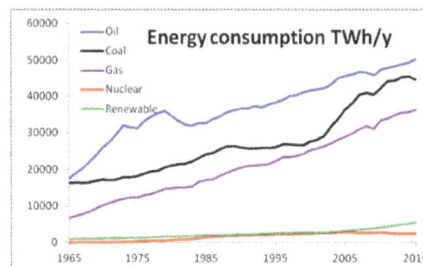

Comparing worldwide energy use, the growth of renewable energy is shown by the green line.

From the end of 2004, worldwide renewable energy capacity grew at rates of 10–60% annually for many technologies. In 2015 global investment in renewables rose 5% to $285.9 billion, breaking the previous record of $278.5 billion in 2011. 2015 was also the first year that saw renewables, excluding large hydro, account for the majority of all new power capacity (134GW, making up 53.6% of the total). Of the renewables total, wind accounted for 72GW and solar photovoltaics 56GW; both record-breaking numbers and sharply up from 2014 figures (49GW and 45GW respectively). In financial terms, solar made up 56% of total new investment and wind accounted for 38%.

Projections vary. The EIA has predicted that almost two thirds of net additions to power capacity will come from renewables by 2020 due to the combined policy benefits of local pollution, decarbonisation and energy diversification. Some studies have set out roadmaps to power 100% of the world's energy with wind, hydroelectric and solar by the year 2030.

According to a 2011 projection by the International Energy Agency, solar power generators may produce most of the world's electricity within 50 years, reducing the emissions of greenhouse gases that harm the environment. Cedric Philibert, senior analyst in the renewable energy division at the IEA said: "Photovoltaic and solar-thermal plants may meet most of the world's demand for electricity by 2060 – and half of all energy needs – with wind, hydropower and biomass plants supplying much of the remaining generation". "Photovoltaic and concentrated solar power together can become the major source of electricity", Philibert said.

In 2014 global wind power capacity expanded 16% to 369,553 MW. Yearly wind energy production is also growing rapidly and has reached around 4% of worldwide electricity usage, 11.4% in the EU, and it is widely used in Asia, and the United States. In 2015, worldwide installed photovoltaics capacity increased to 227 gigawatts (GW), sufficient to supply 1 percent of global electricity demands. Solar thermal energy stations operate in the USA and Spain, and as of 2016, the largest of these is the 392 MW Ivanpah Solar Electric Generating System in California. The world's largest geothermal power installation is The Geysers in California, with a rated capacity of 750 MW. Brazil has one of the largest renewable energy programs in the world, involving production of ethanol fuel from sugar cane, and ethanol now provides 18% of the country's automotive fuel. Ethanol fuel is also widely available in the USA.

Economic Trends

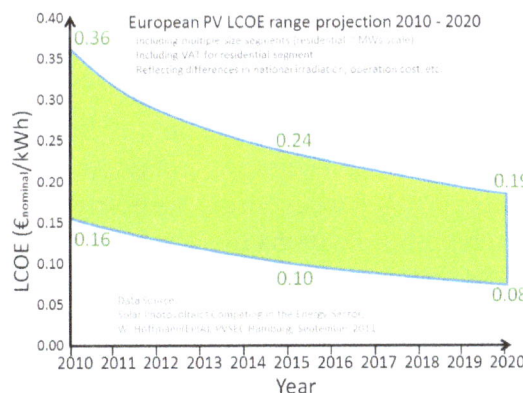

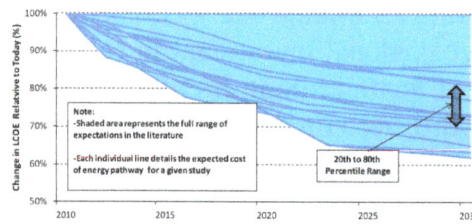

Projection of levelized cost for wind in the U.S. (left) and solar power in Europe.

Renewable energy technologies are getting cheaper, through technological change and through the benefits of mass production and market competition. A 2011 IEA report said: "A portfolio of renewable energy technologies is becoming cost-competitive in an increasingly broad range of circumstances, in some cases providing investment opportunities without the need for specific economic support," and added that "cost reductions in critical technologies, such as wind and solar, are set to continue."

Hydro-electricity and geothermal electricity produced at favourable sites are now the cheapest way to generate electricity. Renewable energy costs continue to drop, and the levelised cost of electricity (LCOE) is declining for wind power, solar photovoltaic (PV), concentrated solar power (CSP) and some biomass technologies. Renewable energy is also the most economic solution for new grid-connected capacity in areas with good resources. As the cost of renewable power falls, the scope of economically viable applications increases. Renewable technologies are now often the most economic solution for new generating capacity. Where "oil-fired generation is the predominant power generation source (e.g. on islands, off-grid and in some countries) a lower-cost renewable solution almost always exists today". A series of studies by the US National Renewable Energy Laboratory modeled the "grid in the Western US under a number of different scenarios where intermittent renewables accounted for 33 percent of the total power." In the models, inefficiencies in cycling the fossil fuel plants to compensate for the variation in solar and wind energy resulted in an additional cost of "between $0.47 and $1.28 to each MegaWatt hour generated"; however, the savings in the cost of the fuels saved "adds up to $7 billion, meaning the added costs are, at most, two percent of the savings."

Hydroelectricity

Only a quarter of the worlds estimated hydroelectric potential of 14,000 TWh/year has been developed, the regional potentials for the growth of hydropower around the world are, 71% Europe, 75% North America, 79% South America, 95% Africa, 95% Middle East, 82% Asia Pacific. However, the political realities of new reservoirs in western countries, economic limitations in the third world and the lack of a transmission system in undeveloped areas, result in the possibility of developing 25% of the remaining potential before 2050, with the bulk of that being in the Asia Pacific area. There is slow growth taking place in Western counties, but not in the conventional dam and reservoir style of the past. New projects take the form of run-of-the-river and small hydro, neither using large reservoirs. It is popular to repower old dams thereby increasing their efficiency and capacity as well as quicker responsiveness on the grid. Where circumstances permit existing dams like the Russell Dam built in 1985 may be updated with "pump back" facilities for pumped-storage which is useful for peak loads or to support intermittent wind and solar power. Countries with large hydroelectric developments like Canada and Norway are spending billions to expand their grids to trade with neighboring countries having limited hydro.

Wind Power Development

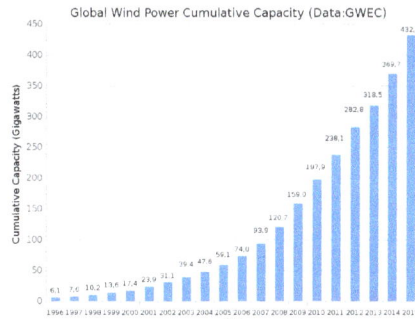

Worldwide growth of wind capacity (1996–2014)

Four offshore wind farms are in the Thames Estuary area: Kentish Flats, Gunfleet Sands,
Thanet and London Array. The latter is the largest in the world as of April 2013.

Wind power is widely used in Europe, China, and the United States. From 2004 to 2014, worldwide installed capacity of wind power has been growing from 47 GW to 369 GW—a more than sevenfold increase within 10 years with 2014 breaking a new record in global installations (51 GW). As of the end of 2014, China, the United States and Germany combined accounted for half of total global capacity. Several other countries have achieved relatively high levels of wind power penetration, such as 21% of stationary electricity production in Denmark, 18% in Portugal, 16% in Spain, and 14% in Ireland in 2010 and have since continued to expand their installed capacity. More than 80 countries around the world are using wind power on a commercial basis.

- Offshore wind power: As of 2014, offshore wind power amounted to 8,771 megawatt of global installed capacity. Although offshore capacity doubled within three years (from 4,117 MW in 2011), it accounted for only 2.3% of the total wind power capacity. The United Kingdom is the undisputed leader of offshore power with half of the world's installed capacity ahead of Denmark, Germany, Belgium and China.

- List of offshore and onshore wind farms: As of 2012, the Alta Wind Energy Center (California, 1,020 MW) is the world's largest wind farm. The London Array (630 MW) is the largest offshore wind farm in the world. The United Kingdom is the world's leading generator of offshore wind power, followed by Denmark. There are several large offshore wind farms operational and under construction and these include Anholt (400 MW), BARD (400 MW), Clyde (548 MW), Fântânele-Cogealac (600 MW), Greater Gabbard (500 MW), Lincs (270 MW), London Array (630 MW), Lower Snake River (343 MW), Macarthur (420 MW), Shepherds Flat (845 MW), and the Sheringham Shoal (317 MW).

Solar Thermal

The 377 MW Ivanpah Solar Electric Generating System with all three towers under load, Feb 2014. Taken from I-15.

Solar Towers of the PS10 and PS20 solar thermal plants in Spain.

The United States conducted much early research in photovoltaics and concentrated solar power. The U.S. is among the top countries in the world in electricity generated by the Sun and several of the world's largest utility-scale installations are located in the desert Southwest.

The oldest solar thermal power plant in the world is the 354 megawatt (MW) SEGS thermal power plant, in California. The Ivanpah Solar Electric Generating System is a solar thermal power project in the California Mojave Desert, 40 miles (64 km) southwest of Las Vegas, with a gross capacity of 377 MW. The 280 MW Solana Generating Station is a solar power plant near Gila Bend, Arizona, about 70 miles (110 km) southwest of Phoenix, completed in 2013. When commissioned it was the largest parabolic trough plant in the world and the first U.S. solar plant with molten salt thermal energy storage.

The solar thermal power industry is growing rapidly with 1.3 GW under construction in 2012 and more planned. Spain is the epicenter of solar thermal power development with 873 MW under construction, and a further 271 MW under development. In the United States, 5,600 MW of solar thermal power projects have been announced. Several power plants have been constructed in the Mojave Desert, Southwestern United States. The Ivanpah Solar Power Facility being the most recent. In developing countries, three World Bank projects for integrated solar thermal/combined-cycle gas-turbine power plants in Egypt, Mexico, and Morocco have been approved.

Photovoltaic Development

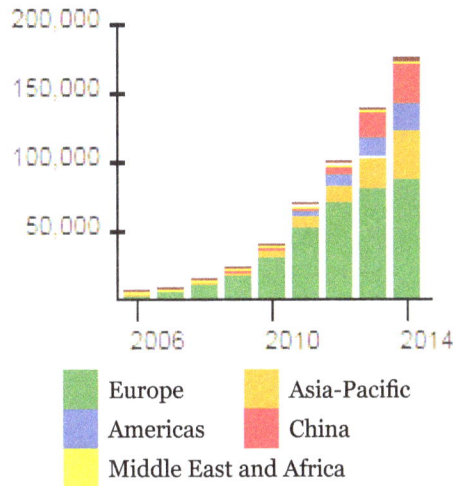

Worldwide growth of PV capacity grouped by region in MW (2006–2014)

Photovoltaics (PV) uses solar cells assembled into solar panels to convert sunlight into electricity. It's a fast-growing technology doubling its worldwide installed capacity every couple of years. PV systems range from small, residential and commercial rooftop or building integrated installations, to large utility-scale photovoltaic power station. The predominant PV technology is crystalline silicon, while thin-film solar cell technology accounts for about 10 percent of global photovoltaic deployment. In recent years, PV technology has improved its electricity generating efficiency, reduced the installation cost per watt as well as its energy payback time, and has reached grid parity in at least 30 different markets by 2014. Financial institutions are predicting a second solar "gold rush" in the near future.

At the end of 2014, worldwide PV capacity reached at least 177,000 megawatts. Photovoltaics grew fastest in China, followed by Japan and the United States, while Germany remains the world's largest overall producer of photovoltaic power, contributing about 7.0 percent to the overall electricity generation. Italy meets 7.9 percent of its electricity demands with photovoltaic power—the highest share worldwide. For 2015, global cumulative capacity is forecasted to increase by more than 50 gigawatts (GW). By 2018, worldwide capacity is projected to reach as much as 430 gigawatts. This corresponds to a tripling within five years. Solar power is forecasted to become the world's largest source of electricity by 2050, with solar photovoltaics and concentrated solar power contributing 16% and 11%, respectively. This requires an increase of installed PV capacity to 4,600 GW, of which more than half is expected to be deployed in China and India.

Photovoltaic Power Stations

Commercial concentrated solar power plants were first developed in the 1980s. As the cost of solar electricity has fallen, the number of grid-connected solar PV systems has grown into the millions and utility-scale solar power stations with hundreds of megawatts are being built. Solar PV is rapidly becoming an inexpensive, low-carbon technology to harness renewable energy from the Sun.

Solar panels at the 550 MW Topaz Solar Farm

Nellis Solar Power Plant, photovoltaic power plant in Nevada, USA.

Many solar photovoltaic power stations have been built, mainly in Europe, China and the USA. The 579 MW Solar Star, in the United States, is the world's largest PV power station.

Many of these plants are integrated with agriculture and some use tracking systems that follow the sun's daily path across the sky to generate more electricity than fixed-mounted systems. There are no fuel costs or emissions during operation of the power stations.

However, when it comes to renewable energy systems and PV, it is not just large systems that matter. Building-integrated photovoltaics or "onsite" PV systems use existing land and structures and generate power close to where it is consumed.

Biofuel Development

Brazil produces bioethanol made from sugarcane available throughout the country. A typical gas station with dual fuel service is marked "A" for alcohol (ethanol) and "G" for gasoline.

Biofuels provided 3% of the world's transport fuel in 2010. Mandates for blending biofuels exist in 31 countries at the national level and in 29 states/provinces. According to the International Energy

Agency, biofuels have the potential to meet more than a quarter of world demand for transportation fuels by 2050.

Since the 1970s, Brazil has had an ethanol fuel program which has allowed the country to become the world's second largest producer of ethanol (after the United States) and the world's largest exporter. Brazil's ethanol fuel program uses modern equipment and cheap sugarcane as feedstock, and the residual cane-waste (bagasse) is used to produce heat and power. There are no longer light vehicles in Brazil running on pure gasoline. By the end of 2008 there were 35,000 filling stations throughout Brazil with at least one ethanol pump. Unfortunately, Operation Car Wash has seriously eroded public trust in oil companies and has implicated several high ranking Brazilian officials.

Nearly all the gasoline sold in the United States today is mixed with 10% ethanol, and motor vehicle manufacturers already produce vehicles designed to run on much higher ethanol blends. Ford, Daimler AG, and GM are among the automobile companies that sell "flexible-fuel" cars, trucks, and minivans that can use gasoline and ethanol blends ranging from pure gasoline up to 85% ethanol. By mid-2006, there were approximately 6 million ethanol compatible vehicles on U.S. roads.

Geothermal Development

Geothermal plant at The Geysers, California, USA.

Geothermal power is cost effective, reliable, sustainable, and environmentally friendly, but has historically been limited to areas near tectonic plate boundaries. Recent technological advances have expanded the range and size of viable resources, especially for applications such as home heating, opening a potential for widespread exploitation. Geothermal wells release greenhouse gases trapped deep within the earth, but these emissions are much lower per energy unit than those of fossil fuels. As a result, geothermal power has the potential to help mitigate global warming if widely deployed in place of fossil fuels.

The International Geothermal Association (IGA) has reported that 10,715 MW of geothermal power in 24 countries is online, which is expected to generate 67,246 GWh of electricity in 2010. This represents a 20% increase in geothermal power online capacity since 2005. IGA projects this will grow to 18,500 MW by 2015, due to the large number of projects presently under consideration, often in areas previously assumed to have little exploitable resource.

In 2010, the United States led the world in geothermal electricity production with 3,086 MW of installed capacity from 77 power plants; the largest group of geothermal power plants in the world is located at The Geysers, a geothermal field in California. The Philippines follows the US as the second highest producer of geothermal power in the world, with 1,904 MW of capacity online; geothermal power makes up approximately 18% of the country's electricity generation.

Developing Countries

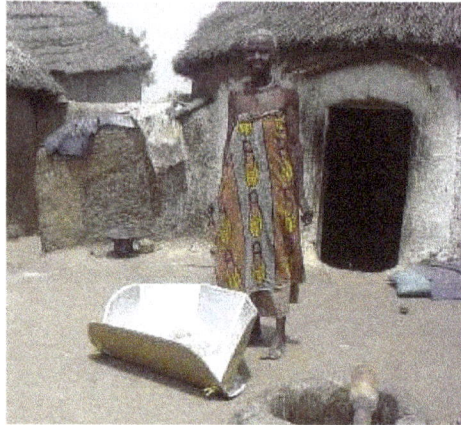

Solar cookers use sunlight as energy source for outdoor cooking.

Renewable energy technology has sometimes been seen as a costly luxury item by critics, and affordable only in the affluent developed world. This erroneous view has persisted for many years, but 2015 was the first year when investment in non-hydro renewables, was higher in developing countries, with $156 billion invested, mainly in China, India, and Brazil.

Renewable energy can be particularly suitable for developing countries. In rural and remote areas, transmission and distribution of energy generated from fossil fuels can be difficult and expensive. Producing renewable energy locally can offer a viable alternative.

Technology advances are opening up a huge new market for solar power: the approximately 1.3 billion people around the world who don't have access to grid electricity. Even though they are typically very poor, these people have to pay far more for lighting than people in rich countries because they use inefficient kerosene lamps. Solar power costs half as much as lighting with kerosene. As of 2010, an estimated 3 million households get power from small solar PV systems. Kenya is the world leader in the number of solar power systems installed per capita. More than 30,000 very small solar panels, each producing 12 to 30 watts, are sold in Kenya annually. Some Small Island Developing States (SIDS) are also turning to solar power to reduce their costs and increase their sustainability.

Micro-hydro configured into mini-grids also provide power. Over 44 million households use biogas made in household-scale digesters for lighting and/or cooking, and more than 166 million households rely on a new generation of more-efficient biomass cookstoves. Clean liquid fuel sourced from renewable feedstocks are used for cooking and lighting in energy-poor areas of the developing world. Alcohol fuels (ethanol and methanol) can be produced sustainably from non-food sugary, starchy, and cellulostic feedstocks. Project Gaia, Inc. and CleanStar Mozambique are implementing clean cooking programs with liquid ethanol stoves in Ethiopia, Kenya, Nigeria and Mozambique.

Renewable energy projects in many developing countries have demonstrated that renewable energy can directly contribute to poverty reduction by providing the energy needed for creating businesses and employment. Renewable energy technologies can also make indirect contributions to alleviating poverty by providing energy for cooking, space heating, and lighting. Renewable energy can also contribute to education, by providing electricity to schools.

Industry and Policy Trends

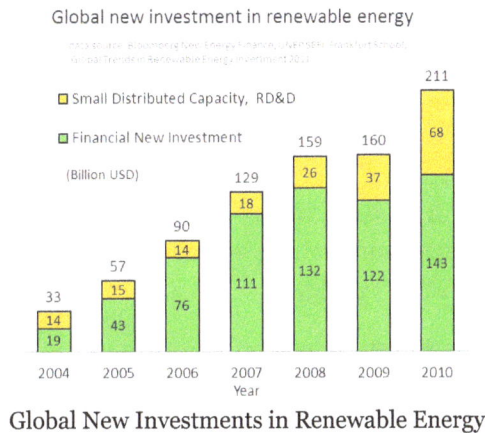

Global New Investments in Renewable Energy

U.S. President Barack Obama's American Recovery and Reinvestment Act of 2009 includes more than $70 billion in direct spending and tax credits for clean energy and associated transportation programs. Leading renewable energy companies include First Solar, Gamesa, GE Energy, Hanwha Q Cells, Sharp Solar, Siemens, SunOpta, Suntech Power, and Vestas.

Many national, state, and local governments have also created green banks. A green bank is a quasi-public financial institution that uses public capital to leverage private investment in clean energy technologies. Green banks use a variety of financial tools to bridge market gaps that hinder the deployment of clean energy.

The military has also focused on the use of renewable fuels for military vehicles. Unlike fossil fuels, renewable fuels can be produced in any country, creating a strategic advantage. The US military has already committed itself to have 50% of its energy consumption come from alternative sources.

The International Renewable Energy Agency (IRENA) is an intergovernmental organization for promoting the adoption of renewable energy worldwide. It aims to provide concrete policy advice and facilitate capacity building and technology transfer. IRENA was formed on 26 January 2009, by 75 countries signing the charter of IRENA. As of March 2010, IRENA has 143 member states who all are considered as founding members, of which 14 have also ratified the statute.

As of 2011, 119 countries have some form of national renewable energy policy target or renewable support policy. National targets now exist in at least 98 countries. There is also a wide range of policies at state/provincial and local levels.

United Nations' Secretary-General Ban Ki-moon has said that renewable energy has the ability to lift the poorest nations to new levels of prosperity. In October 2011, he "announced the creation of a high-level group to drum up support for energy access, energy efficiency and greater use of renewable energy. The group is to be co-chaired by Kandeh Yumkella, the chair of UN Energy and director general of the UN Industrial Development Organisation, and Charles Holliday, chairman of Bank of America".

100% Renewable Energy

The incentive to use 100% renewable energy, for electricity, transport, or even total primary energy supply globally, has been motivated by global warming and other ecological as well as economic concerns. The Intergovernmental Panel on Climate Change has said that there are few fundamental technological limits to integrating a portfolio of renewable energy technologies to meet most of total global energy demand. Renewable energy use has grown much faster than even advocates anticipated. At the national level, at least 30 nations around the world already have renewable energy contributing more than 20% of energy supply. Also, Professors S. Pacala and Robert H. Socolow have developed a series of "stabilization wedges" that can allow us to maintain our quality of life while avoiding catastrophic climate change, and "renewable energy sources," in aggregate, constitute the largest number of their "wedges".

Using 100% renewable energy was first suggested in a Science paper published in 1975 by Danish physicist Bent Sørensen. It was followed by several other proposals, until in 1998 the first detailed analysis of scenarios with very high shares of renewables were published. These were followed by the first detailed 100% scenarios. In 2006 a PhD thesis was published by Czisch in which it was shown that in a 100% renewable scenario energy supply could match demand in every hour of the year in Europa and North Africa. In the same year Danish Energy professor Henrik Lund published a first paper in which he addresses the optimal combination of renewables, which was followed by several other papers on the transition to 100% renewable energy in Denmark. Since then Lund has been publishing several papers on 100% renewable energy. After 2009 publications began to rise steeply, covering 100% scenarios for countries in Europa, America, Australia and other parts of the world.

In 2011 Mark Z. Jacobson, professor of civil and environmental engineering at Stanford University, and Mark Delucchi published a study on 100% renewable global energy supply in the journal Energy Policy. They found producing all new energy with wind power, solar power, and hydropower by 2030 is feasible and existing energy supply arrangements could be replaced by 2050. Barriers to implementing the renewable energy plan are seen to be "primarily social and political, not technological or economic". They also found that energy costs with a wind, solar, water system should be similar to today's energy costs.

Similarly, in the United States, the independent National Research Council has noted that "sufficient domestic renewable resources exist to allow renewable electricity to play a significant role in future electricity generation and thus help confront issues related to climate change, energy security, and the escalation of energy costs ... Renewable energy is an attractive option because renewable resources available in the United States, taken collectively, can supply significantly greater amounts of electricity than the total current or projected domestic demand."

The most significant barriers to the widespread implementation of large-scale renewable energy and low carbon energy strategies are primarily political and not technological. According to the 2013 *Post Carbon Pathways* report, which reviewed many international studies, the key roadblocks are: climate change denial, the fossil fuels lobby, political inaction, unsustainable energy consumption, outdated energy infrastructure, and financial constraints.

Emerging Technologies

Other renewable energy technologies are still under development, and include cellulosic ethanol, hot-dry-rock geothermal power, and marine energy. These technologies are not yet widely demonstrated or have limited commercialization. Many are on the horizon and may have potential comparable to other renewable energy technologies, but still depend on attracting sufficient attention and research, development and demonstration (RD&D) funding.

There are numerous organizations within the academic, federal, and commercial sectors conducting large scale advanced research in the field of renewable energy. This research spans several areas of focus across the renewable energy spectrum. Most of the research is targeted at improving efficiency and increasing overall energy yields. Multiple federally supported research organizations have focused on renewable energy in recent years. Two of the most prominent of these labs are Sandia National Laboratories and the National Renewable Energy Laboratory (NREL), both of which are funded by the United States Department of Energy and supported by various corporate partners. Sandia has a total budget of $2.4 billion while NREL has a budget of $375 million.

- Enhanced geothermal system

  Enhanced geothermal systems (EGS) are a new type of geothermal power technologies that do not require natural convective hydrothermal resources. The vast majority of geothermal energy within drilling reach is in dry and non-porous rock. EGS technologies "enhance" and/or create geothermal resources in this "hot dry rock (HDR)" through hydraulic stimulation. EGS and HDR technologies, like hydrothermal geothermal, are expected to be baseload resources which produce power 24 hours a day like a fossil plant. Distinct from hydrothermal, HDR and EGS may be feasible anywhere in the world, depending on the economic limits of drill depth. Good locations are over deep granite covered by a thick (3–5 km) layer of insulating sediments which slow heat loss. There are HDR and EGS systems currently being developed and tested in France, Australia, Japan, Germany, the U.S. and Switzerland. The largest EGS project in the world is a 25 megawatt demonstration plant currently being developed in the Cooper Basin, Australia. The Cooper Basin has the potential to generate 5,000–10,000 MW.

Enhanced geothermal system

- Cellulosic ethanol

Several refineries that can process biomass and turn it into ethanol are built by companies such as Iogen, POET, and Abengoa, while other companies such as the Verenium Corporation, Novozymes, and Dyadic International are producing enzymes which could enable future commercialization. The shift from food crop feedstocks to waste residues and native grasses offers significant opportunities for a range of players, from farmers to biotechnology firms, and from project developers to investors.

- Marine energy

Rance Tidal Power Station, France

Marine energy (also sometimes referred to as ocean energy) refers to the energy carried by ocean waves, tides, salinity, and ocean temperature differences. The movement of water in the world's oceans creates a vast store of kinetic energy, or energy in motion. This energy can be harnessed to generate electricity to power homes, transport and industries. The term marine energy encompasses both wave power – power from surface waves, and tidal power – obtained from the kinetic energy of large bodies of moving water. Reverse electrodialysis (RED) is a technology for generating electricity by mixing fresh river water and salty sea water in large power cells designed for this purpose; as of 2016 it is being tested at a small scale (50 kW). Offshore wind power is not a form of marine energy, as wind power is derived from the wind, even if the wind turbines are placed over water. The oceans have a tremendous amount of energy and are close to many if not most concentrated populations. Ocean energy has the potential of providing a substantial amount of new renewable energy around the world.

- Experimental solar power

Concentrated photovoltaics (CPV) systems employ sunlight concentrated onto photovoltaic surfaces for the purpose of electricity generation. Thermoelectric, or "thermovoltaic" devices convert a temperature difference between dissimilar materials into an electric current.

- Floating solar arrays

Floating solar arrays are PV systems that float on the surface of drinking water reservoirs, quarry lakes, irrigation canals or remediation and tailing ponds. A small number of such systems exist in France, India, Japan, South Korea, the United Kingdom, Singapore and the United States. The systems are said to have advantages over photovoltaics on land. The cost of land is more expensive, and there are fewer rules and regulations for structures built on bodies of water not used for recreation. Unlike most land-based solar plants, floating arrays can be unobtrusive because they are hidden from public view. They achieve higher efficiencies than PV panels on land, because water cools the panels. The panels have a special coating to prevent rust or corrosion. In May 2008, the Far Niente Winery in Oakville, California, pioneered the world's first floatovoltaic system by installing 994 solar PV modules with a total capacity of 477 kW onto 130 pontoons and floating them on the winery's irrigation pond. Utility-scale floating PV farms are starting to be built. Kyocera will develop the world's largest, a 13.4 MW farm on the reservoir above Yamakura Dam in Chiba Prefecture using 50,000 solar panels. Salt-water resistant floating farms are also being constructed for ocean use. The largest so far announced floatovoltaic project is a 350 MW power station in the Amazon region of Brazil.

- Solar-assisted heat pump

A heat pump is a device that provides heat energy from a source of heat to a destination called a "heat sink". Heat pumps are designed to move thermal energy opposite to the direction of spontaneous heat flow by absorbing heat from a cold space and releasing it to a warmer one. A solar-assisted heat pump represents the integration of a heat pump and thermal solar panels in a single integrated system. Typically these two technologies are used separately (or only placing them in parallel) to produce hot water. In this system the solar thermal panel performs the function of the low temperature heat source and the heat produced is used to feed the heat pump's evaporator. The goal of this system is to get high COP and then produce energy in a more efficient and less expensive way.

It is possible to use any type of solar thermal panel (sheet and tubes, roll-bond, heat pipe, thermal plates) or hybrid (mono/polycrystalline, thin film) in combination with the heat pump. The use of a hybrid panel is preferable because it allows to cover a part of the electricity demand of the heat pump and reduce the power consumption and consequently the variable costs of the system.

- Artificial photosynthesis

Artificial photosynthesis uses techniques including nanotechnology to store solar electromagnetic energy in chemical bonds by splitting water to produce hydrogen and then using carbon dioxide to make methanol. Researchers in this field are striving to design molecular mimics of photosynthesis that utilize a wider region of the solar spectrum, employ catalytic systems made from abundant, inexpensive materials that are robust, readily repaired, non-toxic, stable in a variety of environmental conditions and perform more efficiently allowing a greater proportion of photon energy to end up in the storage compounds, i.e., carbohydrates (rather than building and sustaining living cells). However, prominent research faces hurdles, Sun Catalytix a MIT spin-off stopped scaling up their prototype fuel-cell in 2012, because it offers few savings over other ways to make hydrogen from sunlight.

- Algae fuels

Producing liquid fuels from oil-rich varieties of algae is an ongoing research topic. Various microalgae grown in open or closed systems are being tried including some system that can be set up in brownfield and desert lands.

- Solar aircraft

In 2016, *Solar Impulse 2* was the first solar-powered aircraft to complete circumnavigation of the world.

An electric aircraft is an aircraft that runs on electric motors rather than internal combustion engines, with electricity coming from fuel cells, solar cells, ultracapacitors, power beaming, or batteries.

Currently, flying manned electric aircraft are mostly experimental demonstrators, though many small unmanned aerial vehicles are powered by batteries. Electrically powered model aircraft have been flown since the 1970s, with one report in 1957. The first man-carrying electrically powered flights were made in 1973. Between 2015-2016, a manned, solar-powered plane, Solar Impulse 2, completed a circumnavigation of the Earth.

- Solar updraft tower

The Solar updraft tower is a renewable-energy power plant for generating electricity from low temperature solar heat. Sunshine heats the air beneath a very wide greenhouse-like roofed collector structure surrounding the central base of a very tall chimney tower. The resulting convection causes a hot air updraft in the tower by the chimney effect. This airflow drives wind turbines placed in the chimney updraft or around the chimney base to produce electricity. Plans for scaled-up versions of demonstration models will allow significant power generation, and may allow development of other applications, such as water extraction or distillation, and agriculture or horticulture. A more advanced version of a similarly themed technology is the Vortex engine which aims to replace large physical chimneys with a vortex of air created by a shorter, less-expensive structure.

- Space-based solar power

For either photovoltaic or thermal systems, one option is to loft them into space, particularly Geosynchronous orbit. To be competitive with Earth-based solar power systems, the specific mass (kg/kW) times the cost to loft mass plus the cost of the parts needs to be $2400 or less. I.e., for a parts cost plus rectenna of $1100/kW, the product of the $/kg and kg/kW must be $1300/kW or less. Thus for 6.5 kg/kW, the transport cost cannot exceed $200/kg. While that will require a 100 to one reduction, SpaceX is targeting a ten to one reduction, Reaction Engines may make a 100 to one reduction possible.

Debate

Renewable electricity production, from sources such as wind power and solar power, is sometimes criticized for being variable or intermittent, but is not true for concentrated solar, geothermal and biofuels, that have continuity. In any case, the International Energy Agency has stated that deployment of renewable technologies usually increases the diversity of electricity sources and, through local generation, contributes to the flexibility of the system and its resistance to central shocks.

There have been "not in my back yard" (NIMBY) concerns relating to the visual and other impacts of some wind farms, with local residents sometimes fighting or blocking construction. In the USA, the Massachusetts Cape Wind project was delayed for years partly because of aesthetic concerns. However, residents in other areas have been more positive. According to a town councilor, the overwhelming majority of locals believe that the Ardrossan Wind Farm in Scotland has enhanced the area.

A recent UK Government document states that "projects are generally more likely to succeed if they have broad public support and the consent of local communities. This means giving communities both a say and a stake". In countries such as Germany and Denmark many renewable projects are owned by communities, particularly through cooperative structures, and contribute significantly to overall levels of renewable energy deployment.

The market for renewable energy technologies has continued to grow. Climate change concerns and increasing in green jobs, coupled with high oil prices, peak oil, oil wars, oil spills, promotion of electric vehicles and renewable electricity, nuclear disasters and increasing government support, are driving increasing renewable energy legislation, incentives and commercialization. New government spending, regulation and policies helped the industry weather the 2009 economic crisis better than many other sectors.

Environmental Impact

The ability of biomass and biofuels to contribute to a reduction in CO_2 emissions is limited because both biomass and biofuels emit large amounts of CO_2 when burned. Furthermore, biomass and biofuels consume large amounts of water. Other renewable sources such as wind power, photovoltaics, and hydroelectricity have the advantage of being able to conserve water and reduce CO_2 emissions.

References

- "The surprising history of sustainable energy". Sustainablehistory.wordpress.com. Archived from the original on Dec 24, 2014. Retrieved 1 November 2012

- Vad Mathiesen, Brian; et al. (2015). "Smart Energy Systems for coherent 100% renewable energy and transport solutions". Applied Energy. 145: 139–154. doi:10.1016/j.apenergy.2015.01.075

- Hubbert, M. King (June 1956). "Nuclear Energy and the Fossil Fuels" (PDF). Shell Oil Company/American Petroleum Institute. Retrieved 10 November 2014

- Jacobson, Mark Z.; et al. (2015). "": 100% clean and renewable wind, water, and sunlight (WWS) all-sector energy roadmaps for the 50 United States". Energy and Environmental Science. 8: 2093–2117. doi:10.1039/C5EE01283J

- Armaroli, Nicola; Balzani, Vincenzo (2011). "Towards an electricity-powered world". Energy and Environmental Science. 4: 3193–3222. doi:10.1039/c1ee01249e

- "Infographic: A Timeline Of The Present And Future Of Electric Flight". Popular Science. Retrieved 7 January 2016

- "REN21, Renewables Global Status Report 2012" (PDF). Ren21.net. Archived from the original (PDF) on 11 August 2014. Retrieved 11 August 2014

- Armaroli, Nicola; Balzani, Vincenzo (2016). "Solar Electricity and Solar Fuels: Status and Perspectives in the Context of the Energy Transition". Chemistry – A European Journal. 22: 32–57. doi:10.1002/chem.201503580

- Spellman, Frank R. (2013). Safe Work Practices for Green Energy Jobs (first ed.). DEStech Publications. p. 323. ISBN 978-1-60595-075-4. Retrieved 29 December 2014

- "2014 Outlook: Let the Second Gold Rush Begin" (PDF). Deutsche Bank Markets Research. 6 January 2014. Archived (PDF) from the original on 21 November 2014. Retrieved 22 November 2014

- Demirbas, A. . (2009). "Political, economic and environmental impacts of biofuels: A review". Applied Energy. 86: S108–S117. doi:10.1016/j.apenergy.2009.04.036

Devices used in Energy Harvesting

Wind turbines use wind energy to convert it into electric energy. Small wind turbines, unlike commercial wind turbines, are used for microgenerations. Piezoelectricity, thermoelectric generator and rectenna are devices used in energy harvesting. This chapter helps the readers in developing a better understanding related to the devices used in energy harvesting.

Wind Turbine

A wind turbine is a device that converts the wind's kinetic energy into electrical power.

Wind turbines are manufactured in a wide range of vertical and horizontal axis types. The smallest turbines are used for applications such as battery charging for auxiliary power for boats or caravans or to power traffic warning signs. Slightly larger turbines can be used for making contributions to a domestic power supply while selling unused power back to the utility supplier via the electrical grid. Arrays of large turbines, known as wind farms, are becoming an increasingly important source of intermittent renewable energy and are used by many countries as part of a strategy to reduce their reliance on fossil fuels.

Offshore wind farm, using 5 MW turbines REpower 5M in the North Sea off the coast of Belgium.

History

Wind turbines were used in Persia (present-day Iran) about 500–900 A.D. The windwheel of Hero of Alexandria marks one of the first known instances of wind powering a machine in history. However, the first known practical wind turbines were built in Sistan, an Eastern province of Iran, from the 7th century. These "Panemone" were vertical axle wind turbines, which had long vertical drive shafts with rectangular blades. Made of six to twelve sails covered in reed matting or cloth material, these wind turbines were used to grind grain or draw up water, and were used in the gristmilling and sugarcane industries.

James Blyth's electricity-generating wind turbine, photographed in 1891.

Wind turbines first appeared in Europe during the Middle Ages. The first historical records of their use in England date to the 11th or 12th centuries and there are reports of German crusaders taking their windmill-making skills to Syria around 1190. By the 14th century, Dutch wind turbines were in use to drain areas of the Rhine delta. Advanced wind mills were described by Croatian inventor Fausto Veranzio. In his book Machinae Novae (1595) he described vertical axis wind turbines with curved or V-shaped blades.

The first electricity-generating wind turbine was a battery charging machine installed in July 1887 by Scottish academic James Blyth to light his holiday home in Marykirk, Scotland. Some months later American inventor Charles F. Brush was able to build the first automatically operated wind turbine after consulting local University professors and colleagues Jacob S. Gibbs and Brinsley Coleberd and successfully getting the blueprints peer-reviewed for electricity production in Cleveland, Ohio. Although Blyth's turbine was considered uneconomical in the United Kingdom electricity generation by wind turbines was more cost effective in countries with widely scattered populations.

The first automatically operated wind turbine, built in Cleveland in 1887 by Charles F. Brush. It was 60 feet (18 m) tall, weighed 4 tons (3.6 metric tonnes) and powered a 12 kW generator.

In Denmark by 1900, there were about 2500 windmills for mechanical loads such as pumps and mills, producing an estimated combined peak power of about 30 MW. The largest machines were on 24-meter (79 ft) towers with four-bladed 23-meter (75 ft) diameter rotors. By 1908 there were 72 wind-driven electric generators operating in the United States from 5 kW to 25 kW. Around the time of World War I, American windmill makers were producing 100,000 farm windmills each year, mostly for water-pumping.

By the 1930s, wind generators for electricity were common on farms, mostly in the United States where distribution systems had not yet been installed. In this period, high-tensile steel was cheap, and the generators were placed atop prefabricated open steel lattice towers.

A forerunner of modern horizontal-axis wind generators was in service at Yalta, USSR in 1931. This was a 100 kW generator on a 30-meter (98 ft) tower, connected to the local 6.3 kV distribution system. It was reported to have an annual capacity factor of 32 percent, not much different from current wind machines.

In the autumn of 1941, the first megawatt-class wind turbine was synchronized to a utility grid in Vermont. The Smith-Putnam wind turbine only ran for 1,100 hours before suffering a critical failure. The unit was not repaired, because of shortage of materials during the war.

The first utility grid-connected wind turbine to operate in the UK was built by John Brown & Company in 1951 in the Orkney Islands.

Despite these diverse developments, developments in fossil fuel systems almost entirely eliminated any wind turbine systems larger than supermicro size. In the early 1970s, however, anti-nuclear protests in Denmark spurred artisan mechanics to develop microturbines of 22 kW. Organizing owners into associations and co-operatives lead to the lobbying of the government and utilities and provided incentives for larger turbines throughout the 1980s and later. Local activists in Germany, nascent turbine manufacturers in Spain, and large investors in the United States in the early 1990s then lobbied for policies that stimulated the industry in those countries. Later companies formed in India and China. As of 2012, Danish company Vestas is the world's biggest wind-turbine manufacturer.

Resources

Wind turbine in Germany

A quantitative measure of wind energy available at any location is called the Wind Power Density (WPD). It is a calculation of the mean annual power available per square meter of swept area of a turbine, and is tabulated for different heights above ground. Calculation of wind power density includes the effect of wind velocity and air density. Color-coded maps are prepared for a particular area described, for example, as "Mean Annual Power Density at 50 Metres". In the United States, the results of the above calculation are included in an index developed by the National Renewable Energy Laboratory and referred to as "NREL CLASS". The larger the WPD, the higher it is rated by class. Classes range from Class 1 (200 watts per square meter or less at 50 m altitude) to Class 7 (800 to 2000 watts per square m). Commercial wind farms generally are sited in Class 3 or higher areas, although isolated points in an otherwise Class 1 area may be practical to exploit.

Wind turbines are classified by the wind speed they are designed for, from class I to class IV, with A or B referring to the turbulence.

Class	Avg Wind Speed (m/s)	Turbulence
IA	10	18%
IB	10	16%
IIA	8.5	18%
IIB	8.5	16%
IIIA	7.5	18%
IIIB	7.5	16%
IVA	6	18%
IVB	6	16%

Efficiency

Not all the energy of blowing wind can be used, but some small wind turbines are designed to work at low wind speeds.

Conservation of mass requires that the amount of air entering and exiting a turbine must be equal. Accordingly, Betz's law gives the maximal achievable extraction of wind power by a wind turbine as 16/27 (59.3%) of the total kinetic energy of the air flowing through the turbine.

The maximum theoretical power output of a wind machine is thus 16/27 times the kinetic energy of the air passing through the effective disk area of the machine. If the effective area of the disk is A, and the wind velocity v, the maximum theoretical power output P is:

$$P = \frac{16}{27}\frac{1}{2}\rho v^3 A = \frac{8}{27}\rho v^3 A$$

where ρ is the air density.

As wind is free (no fuel cost), wind-to-rotor efficiency (including rotor blade friction and drag) is one of many aspects impacting the final price of wind power. Further inefficiencies, such as gearbox losses, generator and converter losses, reduce the power delivered by a wind turbine. To protect components from undue wear, extracted power is held constant above the rated operating speed as theoretical power increases at the cube of wind speed, further reducing theoretical efficiency. In 2001, commercial utility-connected turbines deliver 75% to 80% of the Betz limit of power extractable from the wind, at rated operating speed.

Efficiency can decrease slightly over time due to wear. Analysis of 3128 wind turbines older than 10 years in Denmark showed that half of the turbines had no decrease, while the other half saw a production decrease of 1.2% per year. Vertical turbine designs have much lower efficiency than standard horizontal designs.

Types

The three primary types: VAWT Savonius, HAWT towered; VAWT Darrieus as they appear in operation.

Wind turbines can rotate about either a horizontal or a vertical axis, the former being both older and more common. They can also include blades (transparent or not) or be bladeless. Vertical designs produce less power and are less common.

Horizontal Axis

Horizontal-axis wind turbines (HAWT) have the main rotor shaft and electrical generator at the top of a tower, and must be pointed into the wind. Small turbines are pointed by a simple wind vane, while large turbines generally use a wind sensor coupled with a servo motor. Most have a gearbox, which turns the slow rotation of the blades into a quicker rotation that is more suitable to drive an electrical generator.

Components of a horizontal axis wind turbine (gearbox, rotor shaft and brake assembly) being lifted into position.

A turbine blade convoy passing through Edenfield, UK.

Any solid object produces a wake behind it, leading to fatigue failures, so the turbine is usually positioned upwind of its supporting tower. Downwind machines have been built, because they don't need an additional mechanism for keeping them in line with the wind. In high winds, the blades can also be allowed to bend which reduces their swept area and thus their wind resistance. In upwind designs, turbine blades must be made stiff to prevent the blades from being pushed into the tower by high winds. Additionally, the blades are placed a considerable distance in front of the tower and are sometimes tilted forward into the wind a small amount.

Turbines used in wind farms for commercial production of electric power are usually three-bladed. These have low torque ripple, which contributes to good reliability. The blades are usually colored white for daytime visibility by aircraft and range in length from 20 to 80 meters (66 to 262 ft). The size and height of turbines increase year by year. Offshore wind turbines are built up to 8MW today and have a blade length up to 80m. Usual tubular steel towers of multi megawatt turbines have a height of 70 m to 120 m and in extremes up to 160 m.

The blades rotate at 10 to 22 revolutions per minute. At 22 rotations per minute the tip speed exceeds 90 meters per second (300 ft/s). Higher tip speeds means more noise and blade erosion. A gear box is commonly used for stepping up the speed of the generator, although designs may also use direct drive of an annular generator. Some models operate at constant speed, but more energy can be collected by variable-speed turbines which use a solid-state power converter to interface to the transmission system. All turbines are equipped with protective features to avoid damage at high wind speeds, by feathering the blades into the wind which ceases their rotation, supplemented by brakes.

Vertical Axis

Vertical-axis wind turbines (or VAWTs) have the main rotor shaft arranged vertically. One advantage of this arrangement is that the turbine does not need to be pointed into the wind to be effective, which is an advantage on a site where the wind direction is highly variable. It is also an advantage when the turbine is integrated into a building because it is inherently less steerable. Also, the generator and gearbox can be placed near the ground, using a direct drive from the rotor assembly to the ground-based gearbox, improving accessibility for maintenance. However, these designs produce much less energy averaged over time, which is a major drawback.

The key disadvantages include the relatively low rotational speed with the consequential higher torque and hence higher cost of the drive train, the inherently lower power coefficient, the 360-degree rotation of the aerofoil within the wind flow during each cycle and hence the highly dynamic loading on the blade, the pulsating torque generated by some rotor designs on the drive train, and the difficulty of modelling the wind flow accurately and hence the challenges of analysing and designing the rotor prior to fabricating a prototype.

When a turbine is mounted on a rooftop the building generally redirects wind over the roof and this can double the wind speed at the turbine. If the height of a rooftop mounted turbine tower is approximately 50% of the building height it is near the optimum for maximum wind energy and minimum wind turbulence. Wind speeds within the built environment are generally much lower than at exposed rural sites, noise may be a concern and an existing structure may not adequately resist the additional stress.

Subtypes of the vertical axis design include:

Onshore Horizontal Axis Wind Turbines in Zhangjiakou, China.

Offshore Horizontal Axis Wind Turbines (HAWTs) at Scroby Sands Wind Farm, UK.

Darrieus Wind Turbine

"Eggbeater" turbines, or Darrieus turbines, were named after the French inventor, Georges Darrieus. They have good efficiency, but produce large torque ripple and cyclical stress on the tower, which contributes to poor reliability. They also generally require some external power source, or an additional Savonius rotor to start turning, because the starting torque is very low. The torque ripple is reduced by using three or more blades which results in greater solidity of the rotor. Solidity is measured by blade area divided by the rotor area. Newer Darrieus type turbines are not held up by guy-wires but have an external superstructure connected to the top bearing.

Giromill

A subtype of Darrieus turbine with straight, as opposed to curved, blades. The cycloturbine variety has variable pitch to reduce the torque pulsation and is self-starting. The advantages of variable pitch are: high starting torque; a wide, relatively flat torque curve; a higher coefficient of performance; more efficient operation in turbulent winds; and a lower blade speed ratio which lowers blade bending stresses. Straight, V, or curved blades may be used.

Savonius Wind Turbine

These are drag-type devices with two (or more) scoops that are used in anemometers, *Flettner* vents (commonly seen on bus and van roofs), and in some high-reliability low-efficiency power turbines. They are always self-starting if there are at least three scoops.

Twisted Savonius

Twisted Savonius is a modified savonius, with long helical scoops to provide smooth torque. This is often used as a rooftop windturbine and has even been adapted for ships.

Another type of vertical axis is the Parallel turbine, which is similar to the crossflow fan or centrifugal fan. It uses the ground effect. Vertical axis turbines of this type have been tried for many years: a unit producing 10 kW was built by Israeli wind pioneer Bruce Brill in the 1980s.

Design and Construction

Wind turbines are designed, using a range of computer modelling techniques, to exploit the wind energy that exists at a location. For example, Aerodynamic modeling is used to determine the optimum tower height, control systems, number of blades and blade shape.

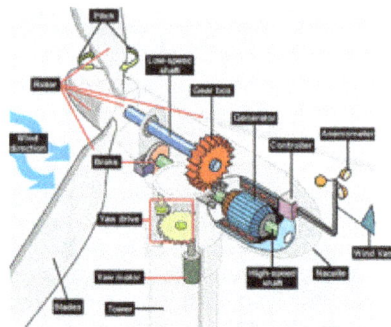

Components of a horizontal-axis wind turbine.

Inside view of a wind turbine tower, showing the tendon cables.

Wind turbines convert wind energy to electricity for distribution. Conventional horizontal axis turbines can be divided into three components:

- The rotor component, which is approximately 20% of the wind turbine cost, includes the blades for converting wind energy to low speed rotational energy.

- The generator component, which is approximately 34% of the wind turbine cost, includes the electrical generator, the control electronics, and most likely a gearbox (e.g. planetary gearbox), adjustable-speed drive or continuously variable transmission component for converting the low speed incoming rotation to high speed rotation suitable for generating electricity.

- The structural support component, which is approximately 15% of the wind turbine cost, includes the tower and rotor yaw mechanism.

A 1.5 MW wind turbine of a type frequently seen in the United States has a tower 80 meters (260 ft) high. The rotor assembly (blades and hub) weighs 22,000 kilograms (48,000 lb). The nacelle, which contains the generator component, weighs 52,000 kilograms (115,000 lb). The concrete base for the tower is constructed using 26,000 kilograms (58,000 lb) of reinforcing steel and contains 190 cubic meters (250 cu yd) of concrete. The base is 15 meters (50 ft) in diameter and 2.4 meters (8 ft) thick near the center.

Among all renewable energy systems wind turbines have the highest effective intensity of power-harvesting surface because turbine blades not only harvest wind power, but also concentrate it.

Unconventional Designs

An E-66 wind turbine in the Windpark Holtriem, Germany, has an observation deck for visitors. Another turbine of the same type with an observation deck is located in Swaffham, England. Airborne wind turbine designs have been proposed and developed for many years but have yet to produce significant amounts of energy. In principle, wind turbines may also be used in conjunction with a large vertical solar updraft tower to extract the energy due to air heated by the sun.

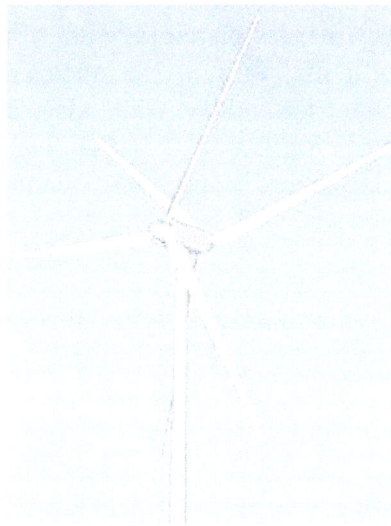

Counter rotating wind turbine (dual rotor)

The corkscrew shaped wind turbine at Progressive Field in Cleveland, Ohio.

Wind turbines which utilise the Magnus effect have been developed.

A ram air turbine (RAT) is a special kind of small turbine that is fitted to some aircraft. When deployed, the RAT is spun by the airstream going past the aircraft and can provide power for the most essential systems if there is a loss of all on-board electrical power, as in the case of the "Gimli Glider".

The two-bladed SCD 6MW offshore turbine designed by aerodyn Energiesysteme and built by MingYang Wind Power has a helideck for helicopters on top of its nacelle. The prototype was erected in 2014 in Rudong, China.

Turbine Monitoring and Diagnostics

Due to data transmission problems, structural health monitoring of wind turbines is usually performed using several accelerometers and strain gages attached to the nacelle to monitor the gearbox and equipments. Currently, digital image correlation and stereophotogrammetry are used to measure dynamics of wind turbine blades. These methods usually measure displacement and strain to identify location of defects. Dynamic characteristics of non-rotating wind turbines have been measured using digital image correlation and photogrammetry. Three dimensional point tracking has also been used to measure rotating dynamics of wind turbines.

Materials and Durability

Materials that are typically used for the rotor blades in wind turbines are composites, as they tend to have a high stiffness, high strength, high fatigue resistance, and low weight. Typical resins used for these composites include polyester and epoxy, while glass and carbon fibers have been used for the reinforcing material. Construction may use manual layup techniques or composite resin injection molding. As the price of glass fibers is only about one tenth the price of carbon fiber, glass fiber is still dominant.

As competition in the wind market increases, companies are seeking ways to draw greater efficiency from their designs. One of the predominant ways wind turbines have gained performance is by increasing rotor diameters, and thus blade length. Retrofitting current turbines with larger blades mitigates the need and risks associated with a system-level redesign. By incorporating carbon fiber into parts of existing blade systems, manufacturers may increase the length of the blades without increasing their overall weight. For instance, the spar cap, a structural element of a turbine blade, commonly experiences high tensile loading, making it an ideal candidate to utilize the enhanced tensile properties of carbon fiber in comparison to glass fiber. Higher stiffness and lower density translates to thinner, lighter blades offering equivalent performance. In a 10-MW turbine—which will become more common in offshore systems by 2021—blade lengths may reach over 100 m and weigh up to 50 metric tons when fabricated out of glass fiber. A switch to carbon fiber in the structural spar of the blade yields weight savings of 20 to 30 percent, or approximately 15 metric tons. The compressive properties of carbon fiber do not differ significantly from those of glass fiber. It is therefore economical to replace glass fiber components under compression with carbon fiber components.

While the material cost is significantly higher for all-glass fiber blades than for hybrid glass/carbon fiber blades, there is a potential for tremendous savings in manufacturing costs when labor price is considered. Utilizing carbon fiber enables for simpler designs that use less raw material. The chief manufacturing process in blade fabrication is the layering of plies. By reducing the number of layers of plies, as is enabled by thinner blade design, the cost of labor may be decreased, and in some cases, equate to the cost of labor for glass fiber blades.

Materials for wind turbine parts other than the rotor blades (including the rotor hub, gearbox, frame, and tower) are largely composed of steel. Modern turbines uses a couple of tonnes of copper for generators, cables and such. Smaller wind turbines have begun incorporating more aluminum based alloys into these components in an effort to make the turbines more lightweight and efficient, and may continue to be used increasingly if fatigue and strength properties can be improved. Prestressed concrete has been increasingly used for the material of the tower, but still, requires much reinforcing steel to meet the strength requirement of the turbine. Additionally, step-up gearboxes are being increasingly replaced with variable speed generators, increasing the demand for magnetic materials in wind turbines., In particular, this would require an increased supply of the rare earth metal neodymium. Reliance on rare earth minerals for components has risked expense and price volatility as China has been main producer of rare earth minerals (96% in 2009) and had been reducing its export quotas of these materials. In recent years, however, other producers have increased production of rare earth minerals and China has removed its reduced export quota on rare earths leading to an increased supply and decreased cost of rare earth minerals, increasing the viability of the implementation of variable speed generators in wind turbines on a large scale.

Wind Turbines on Public Display

The Nordex N50 wind turbine and visitor centre of Lamma Winds in Hong Kong, China.

A few localities have exploited the attention-getting nature of wind turbines by placing them on public display, either with visitor centers around their bases, or with viewing areas farther away. The wind turbines are generally of conventional horizontal-axis, three-bladed design, and generate power to feed electrical grids, but they also serve the unconventional roles of technology demonstration, public relations, and education.

Small Wind Turbines

A small Quietrevolution QR5 Gorlov type vertical axis wind turbine in Bristol, England. Measuring 3 m in diameter and 5 m high, it has a nameplate rating of 6.5 kW to the grid.

Small wind turbines may be used for a variety of applications including on- or off-grid residences, telecom towers, offshore platforms, rural schools and clinics, remote monitoring and other purposes that require energy where there is no electric grid, or where the grid is unstable. Small wind turbines may be as small as a fifty-watt generator for boat or caravan use. Hybrid solar and wind powered units are increasingly being used for traffic signage, particularly in rural locations, as they avoid the need to lay long cables from the nearest mains connection point. The U.S. Department of Energy's National Renewable Energy Laboratory (NREL) defines small wind turbines as those smaller than or equal to 100 kilowatts. Small units often have direct drive generators, direct current output, aeroelastic blades, lifetime bearings and use a vane to point into the wind.

Larger, more costly turbines generally have geared power trains, alternating current output, flaps and are actively pointed into the wind. Direct drive generators and aeroelastic blades for large wind turbines are being researched.

Wind Turbine Spacing

On most horizontal windturbine farms, a spacing of about 6–10 times the rotor diameter is often upheld. However, for large wind farms distances of about 15 rotor diameters should be more economically optimal, taking into account typical wind turbine and land costs. This conclusion has been reached by research conducted by Charles Meneveau of the Johns Hopkins University, and Johan Meyers of Leuven University in Belgium, based on computer simulations that take into account the detailed interactions among wind turbines (wakes) as well as with the entire turbulent atmospheric boundary layer. Moreover, recent research by John Dabiri of Caltech suggests that vertical wind turbines may be placed much more closely together so long as an alternating pattern of rotation is created allowing blades of neighbouring turbines to move in the same direction as they approach one another.

Operability

Maintenance

Wind turbines need regular maintenance to stay reliable and available, reaching 98%.

Modern turbines usually have a small onboard crane for hoisting maintenance tools and minor components. However, large heavy components like generator, gearbox, blades and so on are rarely replaced and a heavy lift external crane is needed in those cases. If the turbine has a difficult access road, a containerized crane can be lifted up by the internal crane to provide heavier lifting.

Repowering

Installation of new wind turbines can be controversial. An alternative is repowering, where existing wind turbines are replaced with bigger, more powerful ones, sometimes in smaller numbers while keeping or increasing capacity.

Demolition

Older turbines were in some early cases not required to be removed when reaching the end of their life. Some still stand, waiting to be recycled or repowered.

A demolition industry develops to recycle offshore turbines at a cost of DKK 2–4 million per MW, to be guaranteed by the owner.

Advantages and Disadvantages

Advantages

Wind turbines are generally inexpensive. They will cost between two and six cents per kilowatt hour, which is one of the lowest-priced renewable energy sources in today's world. And as technology needed for wind turbines continues to improve, the prices will decrease as well. In addition, there is no competitive market for wind energy, as it does not cost money to get ahold of wind. The main cost of wind turbines are the installation process. The average cost is between $48,000 and $65,000 to install. However, the energy harvested from the turbine will offset the installation cost, as well as provide virtually free energy for years after.

Wind turbines provide a clean energy source, emitting no greenhouse gases and no waste product. Over 1,500 tons of carbon dioxide per year can be eliminated by using a one megawatt turbine instead of one megawatt of energy from a fossil fuel. Being environmentally friendly and green is a large advantage of wind turbines.

Wind turbines are also quite efficient. Wind farms can generate between 17 and 39 times as much power as they consume, and in the United States alone, wind turbines have produced about 16 billion kilowatt-hours of energy per year.

Disadvantages

Wind turbines can be very large, reaching over 400 feet tall and with blades 50 yards long, and people have often complained about their visual impact.

Environmental impact of wind power includes effect on wildlife. Thousands of birds, including rare species, have been killed by the blades of wind turbines.

Energy harnessed by wind turbines can be unreliable at times. Wind is often not available when needed, as wind can fluctuate. Turbines can be placed on ridges or bluffs to maximize the access of wind they have, but this also limits the locations where they can be placed. In this way, wind energy is not a particularly reliable source of energy.

Records

Fuhrländer Wind Turbine Laasow, in Brandenburg, Germany, among the world's tallest wind turbines.

Éole, the largest vertical axis wind turbine, in Cap-Chat, Quebec, Canada

Largest Capacity Conventional Drive

The Vestas V164 has a rated capacity of 8 MW, later upgraded to 9 MW. The windmill has an overall height of 220 m (722 ft), a diameter of 164 m (538 ft), is for offshore use, and is the world's largest-capacity wind turbine since its introduction in 2014. The conventional drive train consist of a main gearbox and a medium speed PM generator. Prototype installed in 2014 at the National Test Center Denmark nearby Østerild. Series production began end of 2015.

Largest Capacity Direct Drive

The Enercon E-126 with 7.58 MW and 127 m rotor diameter is the largest direct drive turbine. It's only for onshore use. The turbine has parted rotor blades with 2 sections for transport. In July 2016, Siemens upgraded its 7 to 8 MW.

Largest Vertical-axis

Le Nordais wind farm in Cap-Chat, Quebec has a vertical axis wind turbine (VAWT) named Éole, which is the world's largest at 110 m. It has a nameplate capacity of 3.8 MW.

Largest 1-bladed Turbine

Riva Calzoni M33 was a single-bladed wind turbine with 350 kW, designed and built In Bologna in 1993.

Largest 2-bladed Turbine

The biggest 2-bladed turbine is built by Mingyang Wind Power in 2013. It is a SCD6.5MW offshore downwind turbine, designed by aerodyn Energiesysteme.

Largest Swept Area

The turbine with the largest swept area is the Samsung S7.0–171, with a diameter of 171 m, giving a total sweep of 22966 m^2.

Tallest

A Nordex 3.3 MW was installed in July 2016. It has a total height of 230m, and a hub height of 164m on 100m concrete tower bottom with steel tubes on top (hybrid tower).

Vestas V164 was the tallest wind turbine, standing in Østerild, Denmark, 220 meters tall, constructed in 2014. It has a steel tube tower.

Highest Tower

Fuhrländer installed a 2.5MW turbine on a 160m lattice tower in 2003.

Most Rotors

Lagerwey has build Four-in-One, a multi rotor wind turbine with one tower and four rotors near Maasvlakte. In April 2016, Vestas installed a 900 kW quadrotor test wind turbine at Risø, made from 4 recycled 225 kW V29 turbines.

Most Productive

Four turbines at Rønland Offshore Wind Farm in Denmark share the record for the most productive wind turbines, with each having generated 63.2 GWh by June 2010.

Highest-situated

Since 2013 the world's highest-situated wind turbine was made and installed by WindAid and is located at the base of the Pastoruri Glacier in Peru at 4,877 meters (16,001 ft) above sea level. The site uses the WindAid 2.5 kW wind generator to supply power to a small rural community of micro entrepreneurs who cater to the tourists who come to the Pastoruri glacier.

Largest Floating Wind Turbine

The world's largest—and also the first operational deep-water *large-capacity*—floating wind turbine is the 2.3 MW Hywind currently operating 10 kilometers (6.2 mi) offshore in 220-meter-deep water, southwest of Karmøy, Norway. The turbine began operating in September 2009 and utilizes a Siemens 2.3 MW turbine.

Small Wind Turbine

A group of small wind turbines in a community in Dali, Yunnan, China.

A small wind turbine is a wind turbine used for microgeneration, as opposed to large commercial wind turbines, such as those found in wind farms, with greater individual power output. The Canadian Wind Energy Association (CanWEA) defines "small wind" as ranging from less than 1000 Watt (1 kW) turbines up to 300 kW turbines. The smaller turbines may be as small as a 50 Watt auxiliary power generator for a boat, caravan, or miniature refrigeration unit.

Design

Smaller scale turbines for residential scale use are available. Their blades are usually 1.5 to 3.5 metres (4 ft 11 in–11 ft 6 in) in diameter and produce 1-10 kW of electricity at their optimal wind speed. Some units have been designed to be very lightweight in their construction, e.g. 16 kilograms (35 lb), allowing sensitivity to minor wind movements and a rapid response to wind gusts typically found in urban settings and easy mounting much like a television antenna. It is claimed, and a few are certified, as being inaudible even a few feet (about a metre) under the turbine.

The majority of small wind turbines are traditional horizontal axis wind turbines, but vertical axis wind turbines are a growing type of wind turbine in the small-wind market. Makers of vertical axis wind turbines such as WePower, Urban Green Energy, Helix Wind, and Windspire Energy, have reported increasing sales over the previous years.

Evance R9000 5kW small domestic wind turbine with 5.5m rotor diameter.

The generators for small wind turbines usually are three-phase alternating current generators and the trend is to use the induction type. They are options for direct current output for battery charging and power inverters to convert the power back to AC but at constant frequency for grid connectivity. Some models utilize single-phase generators.

Some small wind turbines can be designed to work at low wind speeds, but in general small wind turbines require a minimum wind speed of 4 metres per second (13 ft/s).

Dynamic braking regulates the speed by dumping excess energy, so that the turbine continues to produce electricity even in high winds. The dynamic braking resistor may be installed inside the building to provide heat (during high winds when more heat is lost by the building, while more heat is also produced by the braking resistor). The location makes low voltage (around 12 volt) distribution practical.

Small units often have direct drive generators, direct current output, lifetime bearings and use a vane to point into the wind. Larger, more costly turbines generally have geared power trains, alternating current output and are actively pointed into the wind. Direct drive generators are also used on some large wind turbines.

Installation

Turbines are often mounted on a tower to raise them above any nearby obstacles. One rule of thumb is that turbines should be at least 30 ft (9.1 m) higher than anything within 500 ft (150 m). Better locations for wind turbines are far away from large upwind obstacles. Measurements made in a boundary layer wind tunnel have indicated that significant detrimental effects associated with nearby obstacles can extend up to 80 times the obstacle's height downwind. However, this is an extreme case. Another approach to siting a small turbine is to use a shelter model to predict how nearby obstacles will affect local wind conditions. Models of this type are general and can be applied to any site. They are often developed based on actual wind measurements, and can estimate flow properties such as mean wind speed and turbulence levels at a potential turbine location, taking into account the size, shape, and distance to any nearby obstacles.

Twisted savonius wind turbine in Granville, France

A small wind turbine can be installed on a roof. Installation issues then include the strength of the roof, vibration, and the turbulence caused by the roof ledge. Small-scale rooftop turbines suffer from turbulence and rarely generate significant amounts of power, especially in towns and cities.

Markets

Japan

In July 2012, a new feed-in tariff approved by Japanese Industry Minister Yukio Edano went into effect, promising to boost the country's production of wind and solar energy production. The country is aiming to increase renewable energy investment in part as a response to the Fukushima radiation crisis in March 2011. The feed-in tariff applies to solar panels and small wind turbines and requires utilities to buy back electricity generated from renewable energy sources at government-established rates.

Small-scale wind power (turbines of less than 20 kW capacity) will be subsidized at least 57.75 JPY (about 0.74 USD per kwh).

United Kingdom

Properties in rural or suburban parts of the UK can opt for a wind turbine with inverter to supplement local grid power. The UK's Microgeneration Certification Scheme (MCS) provides feed-in tariffs to owners of qualified small wind turbines.

United States

Small wind turbines added a total of 17.3 MW of generating capacity throughout the United States in 2008, according to the American Wind Energy Association (AWEA). That growth equaled a 78% increase in the domestic market for small wind turbines, which are defined as wind turbines with capacities of 100 kW or less. AWEA's "2009 Small Wind Global Market Study", published in late 2009 May, credited the increase in part to greater manufacturing volumes, as the industry was able to attract enough private investment to finance manufacturing plant expansions. It also credited rising electricity prices and greater public awareness of wind technologies for an increase in residential sale. But a poll of small wind manufacturers found that the growth in 2008 might be only a glimmer of things to come, as the companies projected a 30-fold growth in the U.S. small wind market within as little as five years, despite the global recession. The U.S. small wind industry also benefits from the global market, as it controls about half of the global market share. U.S. manufacturers garnered $77 million of the $156 million that was spent throughout the world on small wind turbine installations. A total of 38.7 MW of small wind power capacity was installed globally in 2008.

In the United States, residential wind turbines with outputs of 2–10 kW, typically cost between US$12,000 and US$55,000 installed (US$6 per watt), although there are incentives and rebates available in 19 states that can reduce the purchase price for homeowners by up to 50 percent, to $3 per watt. The US manufacturer Southwest Windpower estimates a turbine to pay for itself in energy savings in 5 to 12 years.

The dominant models on the market, especially in the United States, are horizontal-axis wind turbines.

To enable consumers to make an informed decision when purchasing a small wind turbine, a method for consumer labeling has been developed by IEA Wind Task 27 in collaboration with IEC TC88 MT2. In 2011 IEA Wind published a Recommended Practice, which describes the tests and procedures required to apply the label.

Croatia

Hybrid system, 2400W windturbines, 4000W solar modules, island Žirje, Croatia.

Croatia is an ideal market for small wind turbines due to Mediterranean climate and numerous islands with no access to the electric grid. In winter months when there is less sun, but more wind, small wind turbines are a great addition to isolated renewable energy sites (GSM, stations, marinas etc.). That way solar and wind power provide consistent energy throughout the year.

Germany

In Germany the feed-in tariff for small wind turbines has always been the same as for large turbines. This is the main reason the small wind turbine sector in Germany developed slowly. In contrast, small photovoltaic systems in Germany benefited from a high feed-in tariff, at times above 50 Euro-Cent per kilowatt hour.

In August 2014 the German renewable energy law was adjusted, also affecting the feed-in tariffs for wind turbines. For the operation of a small wind turbine with a capacity below 50 kilowatt the tariff amounts to 8.5 Euro-Cent for a period of 20 years.

Due to the low feed-in tariff and high electricity prices in Germany, the economic operation of a small wind turbine depends on a large self-consumption rate of the electricity produced by the small wind turbine. Private households pay on average 28 cent per kilowatt hour for electricity (19% VAT included).

As part of the German renewable energy law 2014 a fee on self-consumed electricity was introduced in August 2014. The regulation does not apply to small power plants with a capacity below 10 kilowatt. With an amount of 1.87 Euro-Cents the fee is low.

DIY Construction

Some hobbyists have built wind turbines from kits, sourced components, or from scratch. DIY wind turbines are usually smaller (rooftop) turbines of approximately 1 kW or less. These small wind turbines are usually tilt-up or fixed / guyed towers.

Do it yourself or DIY-wind turbine construction has been made popular by magazines such as OtherPower and Home Power.

Organizations as Practical Action have designed DIY wind turbines that can be easily built by communities in developing nations and are supplying concrete documents on how to do so.

Wind Turbine Design

An example of a wind turbine, this 3 bladed turbine is the classic design of modern wind turbines.

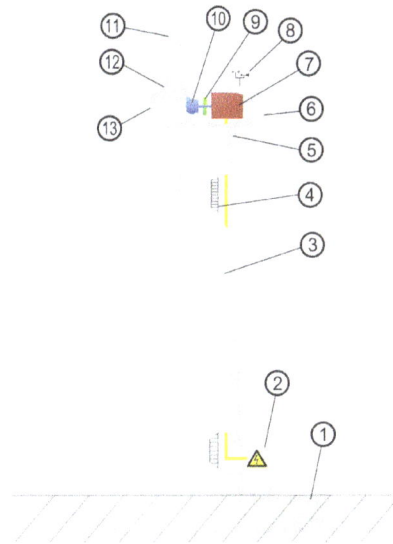

Wind turbine components : 1-Foundation, 2-Connection to the electric grid, 3-Tower, 4-Access ladder, 5-Wind orientation control (Yaw control), 6-Nacelle, 7-Generator, 8-Anemometer, 9-Electric or Mechanical Brake, 10-Gearbox, 11-Rotor blade, 12-Blade pitch control, 13-Rotor hub.

Wind turbine design is the process of defining the form and specifications of a wind turbine to extract energy from the wind. A wind turbine installation consists of the necessary systems needed to capture the wind's energy, point the turbine into the wind, convert mechanical rotation into electrical power, and other systems to start, stop, and control the turbine.

This section covers the design of horizontal axis wind turbines (HAWT) since the majority of commercial turbines use this design.

In 1919 the physicist Albert Betz showed that for a hypothetical ideal wind-energy extraction machine, the fundamental laws of conservation of mass and energy allowed no more than 16/27 (59.3%) of the kinetic energy of the wind to be captured. This Betz' law limit can be approached by modern turbine designs which may reach 70 to 80% of this theoretical limit.

In addition to aerodynamic design of the blades, design of a complete wind power system must also address design of the hub, controls, generator, supporting structure and foundation. Further design questions arise when integrating wind turbines into electrical power grids.

Aerodynamics

The shape and dimensions of the blades of the wind turbine are determined by the aerodynamic performance required to efficiently extract energy from the wind, and by the strength required to resist the forces on the blade.

The aerodynamics of a horizontal-axis wind turbine are not straightforward. The air flow at the blades is not the same as the airflow far away from the turbine. The very nature of the way in which energy is extracted from the air also causes air to be deflected by the turbine. In addition the aerodynamics of a wind turbine at the rotor surface exhibit phenomena that are rarely seen in other aerodynamic fields.

Wind rotor profile

In 1919 the physicist Albert Betz showed that for a hypothetical ideal wind-energy extraction machine, the fundamental laws of conservation of mass and energy allowed no more than 16/27 (59.3%) of the kinetic energy of the wind to be captured. This Betz' law limit can be approached by modern turbine designs which may reach 70 to 80% of this theoretical limit.

Power Control

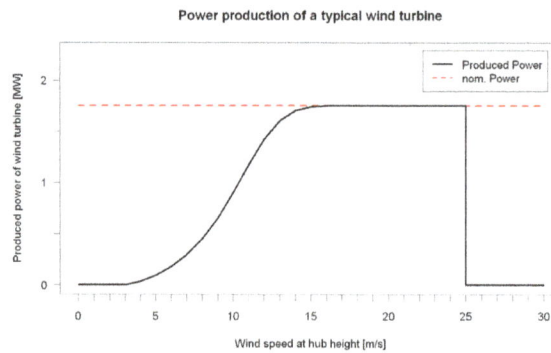

The speed at which a wind turbine rotates must be controlled for efficient power generation and to keep the turbine components within designed speed and torque limits. The centrifugal force on the spinning blades increases as the square of the rotation speed, which makes this structure sensitive to overspeed. Because the power of the wind increases as the cube of the wind speed, turbines have to be built to survive much higher wind loads (such as gusts of wind) than those from which they can practically generate power. Wind turbines have ways of reducing torque in high winds.

A wind turbine is designed to produce power over a range of wind speeds. The cut-in speed is around 3-4 m/s for most turbines, and cut-out at 25 m/s. If the rated wind speed is exceeded the power has to be limited. There are various ways to achieve this.

A control system involves three basic elements: sensors to measure process variables, actuators to manipulate energy capture and component loading, and control algorithms to coordinate the actuators based on information gathered by the sensors.

All wind turbines are designed for a maximum wind speed, called the survival speed, above which they will be damaged. The survival speed of commercial wind turbines is in the range of 40 m/s (144 km/h, 89 MPH) to 72 m/s (259 km/h, 161 MPH). The most common survival speed is 60 m/s (216 km/h, 134 MPH).

Stall

Stalling works by increasing the angle at which the relative wind strikes the blades (angle of attack), and it reduces the induced drag (drag associated with lift). Stalling is simple because it can be made to happen passively (it increases automatically when the winds speed up), but it increases the cross-section of the blade face-on to the wind, and thus the ordinary drag. A fully stalled turbine blade, when stopped, has the flat side of the blade facing directly into the wind.

A fixed-speed HAWT (Horizontal Axis Wind Turbine) inherently increases its angle of attack at higher wind speed as the blades speed up. A natural strategy, then, is to allow the blade to stall when the wind speed increases. This technique was successfully used on many early HAWTs. However, on some of these blade sets, it was observed that the degree of blade pitch tended to increase audible noise levels.

Vortex generators may be used to control the lift characteristics of the blade. The VGs are placed on the airfoil to enhance the lift if they are placed on the lower (flatter) surface or limit the maximum lift if placed on the upper (higher camber) surface.

Furling works by decreasing the angle of attack, which reduces the induced drag from the lift of the rotor, as well as the cross-section. One major problem in designing wind turbines is getting the blades to stall or furl quickly enough should a gust of wind cause sudden acceleration. A fully furled turbine blade, when stopped, has the edge of the blade facing into the wind.

Loads can be reduced by making a structural system softer or more flexible. This could be accomplished with downwind rotors or with curved blades that twist naturally to reduce angle of attack at higher wind speeds. These systems will be nonlinear and will couple the structure to the flow field - thus, design tools must evolve to model these nonlinearities.

Standard modern turbines all furl the blades in high winds. Since furling requires acting against the torque on the blade, it requires some form of pitch angle control, which is achieved with a slewing drive. This drive precisely angles the blade while withstanding high torque loads. In addition, many turbines use hydraulic systems. These systems are usually spring-loaded, so that if hydraulic power fails, the blades automatically furl. Other turbines use an electric servomotor for every rotor blade. They have a small battery-reserve in case of an electric-grid breakdown. Small wind turbines (under 50 kW) with variable-pitching generally use systems operated by centrifugal force, either by flyweights or geometric design, and employ no electric or hydraulic controls.

Fundamental gaps exist in pitch control, limiting the reduction of energy costs, according to a report from a coalition of researchers from universities, industry, and government, supported by the Atkinson Center for a Sustainable Future. Load reduction is currently focused on full-span blade pitch control, since individual pitch motors are the actuators currently available on commercial turbines. Significant load mitigation has been demonstrated in simulations for blades, tower, and drive train. However, there is still research needed, the methods for realization of full-span blade pitch control need to be developed in order to increase energy capture and mitigate fatigue loads.

A control technique applied to the pitch angle is done by comparing the current active power of the engine with the value of active power at the rated engine speed (active power reference, Ps reference). Control of the pitch angle in this case is done with a PI controller controls. However, in order to have a realistic response to the control system of the pitch angle, the actuator uses the time constant Tservo, an integrator and limiters so as the pitch angle to be from 0° to 30° with a rate of change (± 10° per sec).

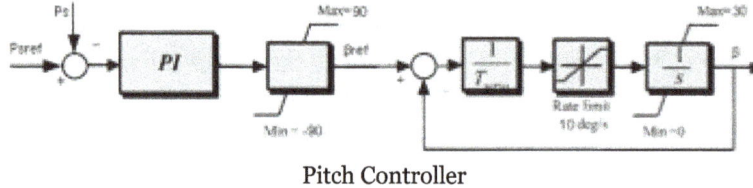

Pitch Controller

From the figure, the reference pitch angle is compared with the actual pitch angle b and then the error is corrected by the actuator. The reference pitch angle, which comes from the PI controller, goes through a limiter. Restrictions on limits are very important to maintain the pitch angle in real term. Limiting the rate of change is very important especially during faults in the network. The importance is due to the fact that the controller decides how quickly it can reduce the aerodynamic energy to avoid acceleration during errors.

Other Controls

Generator Torque

Modern large wind turbines are variable-speed machines. When the wind speed is below rated, generator torque is used to control the rotor speed in order to capture as much power as possible. The most power is captured when the tip speed ratio is held constant at its optimum value (typically 6 or 7). This means that as wind speed increases, rotor speed should increase proportionally. The difference between the aerodynamic torque captured by the blades and the applied generator torque controls the rotor speed. If the generator torque is lower, the rotor accelerates, and if the generator torque is higher, the rotor slows down. Below rated wind speed, the generator torque control is active while the blade pitch is typically held at the constant angle that captures the most power, fairly flat to the wind. Above rated wind speed, the generator torque is typically held constant while the blade pitch is active.

One technique to control a permanent magnet synchronous motor is Field Oriented Control. Field Oriented Control is a closed loop strategy composed of two current controllers (an inner loop and outer loop cascade design) necessary for controlling the torque, and one speed controller.

Constant torque angle control

In this control strategy the d axis current is kept zero, while the vector current is align with the q axis in order to maintain the torque angle equal with 90°. This is one of the most used control strategy because of the simplicity, by controlling only the Iqs current. So, now the electromagnetic torque equation of the permanent magnet synchronous generator is simply a linear equation depend on the Iqs current only.

So, the electromagnetic torque for Ids = 0 (we can achieve that with the d-axis controller) is now:

$$T_e = 3/2 \ p \ (\lambda_{pm} I_{qs} + (L_{ds} - L_{qs}) I_{ds} I_{qs}) = 3/2 \ p \ \lambda_{pm} I_{qs}$$

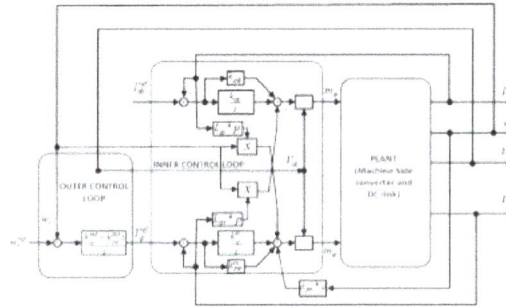

Machine Side Controller Design

So, the complete system of the machine side converter and the cascaded PI controller loops is given by the figure. In that we have the control inputs, which are the duty rations m_{ds} and m_{qs}, of the PWM-regulated converter. Also, we can see the control scheme for the wind turbine in the machine side and simultaneously how we keep the I_{ds} zero (the electromagnetic torque equation is linear).

Yawing

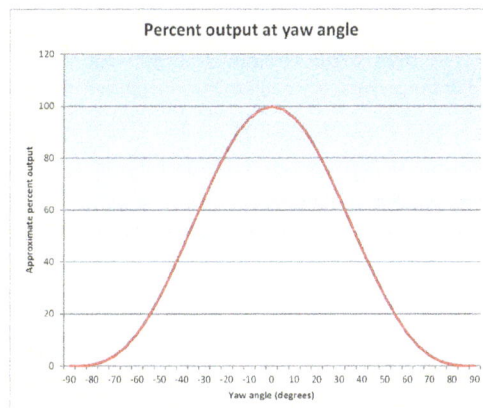

Percent output vs. wind angle

Modern large wind turbines are typically actively controlled to face the wind direction measured by a wind vane situated on the back of the nacelle. By minimizing the yaw angle (the misalignment between wind and turbine pointing direction), the power output is maximized and non-symmetrical loads minimized. However, since the wind direction varies quickly the turbine will not strictly follow the direction and will have a small yaw angle on average. The power output losses can simply be approximated to fall with (cos(yaw angle))³. Particularly at low-to-medium wind speeds, yawing can make a significant reduction in turbine output, with wind direction variations of ±30° being quite common and long response times of the turbines to changes in wind direction. At high wind speeds, the wind direction is less variable.

Electrical Braking

Braking of a small wind turbine can be done by dumping energy from the generator into a resistor bank, converting the kinetic energy of the turbine rotation into heat. This method is useful if the kinetic load on the generator is suddenly reduced or is too small to keep the turbine speed within its allowed limit.

Cyclically braking causes the blades to slow down, which increases the stalling effect, reducing the efficiency of the blades. This way, the turbine's rotation can be kept at a safe speed in faster winds while maintaining (nominal) power output. This method is usually not applied on large grid-connected wind turbines.

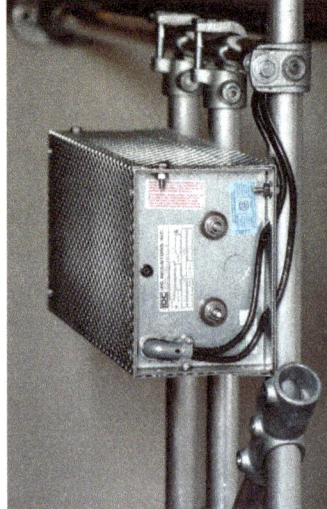

2kW Dynamic braking resistor for small wind turbine.

Mechanical Braking

A mechanical drum brake or disk brake is used to stop turbine in emergency situation such as extreme gust events or over speed. This brake is a secondary means to hold the turbine at rest for maintenance, with a rotor lock system as primary means. Such brakes are usually applied only after blade furling and electromagnetic braking have reduced the turbine speed generally 1 or 2 rotor RPM, as the mechanical brakes can create a fire inside the nacelle if used to stop the turbine from full speed. The load on the turbine increases if the brake is applied at rated RPM. Mechanical brakes are driven by hydraulic systems and are connected to main control box.

Turbine Size

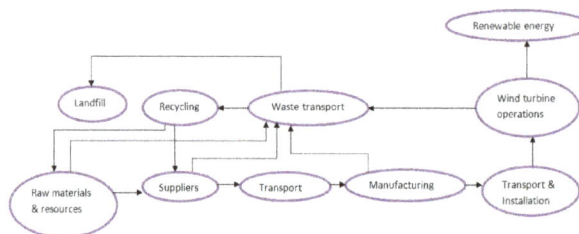

Flow diagram for wind turbine plant

There are different size classes of wind turbines. The smallest having power production less than 10 kW are used in homes, farms and remote applications whereas intermediate wind turbines (10-250 kW) are useful for village power, hybrid systems and distributed power. The world's largest wind turbine, an 8-MW turbine located at the Burbo Bank Extension wind farm in Liverpool Bay, United Kingdom, was installed in 2016. Utility-scale turbines (larger than one megawatt) are used in central station wind farms, distributed power and community wind.

A person standing beside 15 m long blades

For a given survivable wind speed, the mass of a turbine is approximately proportional to the cube of its blade-length. Wind power intercepted by the turbine is proportional to the square of its blade-length. The maximum blade-length of a turbine is limited by both the strength and stiffness of its material.

Labor and maintenance costs increase only gradually with increasing turbine size, so to minimize costs, wind farm turbines are basically limited by the strength of materials, and siting requirements.

Typical modern wind turbines have diameters of 40 to 90 metres (130 to 300 ft) and are rated between 500 kW and 2 MW. As of 2014 the most powerful turbine, the Vestas V-164, is rated at 8 MW and has a rotor diameter of 164m.

Nacelle

The nacelle is housing the gearbox and generator connecting the tower and rotor. Sensors detect the wind speed and direction, and motors turn the nacelle into the wind to maximize output.

Gearbox

In conventional wind turbines, the blades spin a shaft that is connected through a gearbox to the generator. The gearbox converts the turning speed of the blades 15 to 20 rotations per minute for a large, one-megawatt turbine into the faster 1,800 revolutions per minute that the generator needs to generate electricity. Analysts from GlobalData estimate that gearbox market grows from \$3.2bn in 2006 to \$6.9bn in 2011, and to \$8.1bn by 2020. Market leaders were Winergy in 2011. The use of magnetic gearboxes has also been explored as a way of reducing wind turbine maintenance costs.

Generator

For large, commercial size horizontal-axis wind turbines, the electrical generator is mounted in a nacelle at the top of a tower, behind the hub of the turbine rotor. Typically wind turbines generate electricity through asynchronous machines that are directly connected with the electricity grid. Usually the rotational speed of the wind turbine is slower than the equivalent rotation speed of the electrical network: typical rotation speeds for wind generators are 5–20 rpm while a directly connected machine will have an electrical speed between 750 and 3600 rpm. Therefore, a gearbox is inserted between the rotor hub and the generator. This also reduces the generator cost and weight. Commercial size generators have a rotor carrying a field winding so that a rotating magnetic field is produced inside a set of windings called the stator. While the rotating field winding consumes a fraction of a percent of the generator output, adjustment of the field current allows good control over the generator output voltage.

Gearbox, rotor shaft and brake assembly

Older style wind generators rotate at a constant speed, to match power line frequency, which allowed the use of less costly induction generators. Newer wind turbines often turn at whatever speed generates electricity most efficiently. The varying output frequency and voltage can be matched to the fixed values of the grid using multiple technologies such as doubly fed induction generators or full-effect converters where the variable frequency current produced is converted to DC and then back to AC. Although such alternatives require costly equipment and cause power loss, the turbine can capture a significantly larger fraction of the wind energy. In some cases, especially when turbines are sited offshore, the DC energy will be transmitted from the turbine to a central (onshore) inverter for connection to the grid.

Gearless Wind Turbine

Gearless wind turbines (also called direct drive) get rid of the gearbox completely. Instead, the rotor shaft is attached directly to the generator, which spins at the same speed as the blades. Enercon and EWT (formerly known as Lagerwey) have produced gearless wind turbines with separately electrically excited generators for many years, and Siemens produces a gearless "inverted generator" 3 MW model while developing a 6 MW model. To make up for a direct drive generator's slower spinning rate, the diameter of the generator's rotor is increased so that it can contain more magnets to create the required frequency and power.

Gearless wind turbines are often heavier than gear based wind turbines. A study by the EU called "Reliawind" based on the largest sample size of turbines has shown that the reliability of gearboxes is not the main problem in wind turbines. The reliability of direct drive turbines offshore is still not known, since the sample size is so small.

Experts from Technical University of Denmark estimate that a geared generator with permanent magnets may use 25 kg/MW of the rare earth element Neodymium, while a gearless may use 250 kg/MW.

In December 2011, the US Department of Energy published a report stating critical shortage of rare earth elements such as neodymium used in large quantities for permanent magnets in gearless wind turbines. China produces more than 95% of rare earth elements, while Hitachi holds more than 600 patents covering Neodymium magnets. Direct-drive turbines require 600 kg of permanent magnet material per megawatt, which translates to several hundred kilograms of rare earth content per megawatt, as neodymium content is estimated to be 31% of magnet weight. Hybrid drivetrains (intermediate between direct drive and traditional geared) use significantly less rare earth materials. While permanent magnet wind turbines only account for about 5% of the market outside of China, their market share inside of China is estimated at 25% or higher. In 2011, demand for neodymium in wind turbines was estimated to be 1/5 of that in electric vehicles.

Blades

Blade Design

Unpainted tip of a blade

The ratio between the speed of the blade tips and the speed of the wind is called tip speed ratio. High efficiency 3-blade-turbines have tip speed/wind speed ratios of 6 to 7. Modern wind turbines are designed to spin at varying speeds (a consequence of their generator design). Use of aluminum and composite materials in their blades has contributed to low rotational inertia, which means that newer wind turbines can accelerate quickly if the winds pick up, keeping the tip speed ratio more nearly constant. Operating closer to their optimal tip speed ratio during energetic gusts of wind allows wind turbines to improve energy capture from sudden gusts that are typical in urban settings.

And in contrast, older style wind turbines were designed with heavier steel blades, which have higher inertia, and rotated at speeds governed by the AC frequency of the power lines. The high inertia buffered the changes in rotation speed and thus made power output more stable.

It is generally understood that noise increases with higher blade tip speeds. To increase tip speed without increasing noise would allow reduction the torque into the gearbox and generator and reduce overall structural loads, thereby reducing cost. The reduction of noise is linked to the detailed aerodynamics of the blades, especially factors that reduce abrupt stalling. The inability to predict stall restricts the development of aggressive aerodynamic concepts. Some blades (mostly on Enercon) have a winglet to increase performance and/or reduce noise.

A blade can have a lift-to-drag ratio of 120, compared to 70 for a sailplane and 15 for an airliner.

The Hub

A Wind turbine hub being installed

In simple designs, the blades are directly bolted to the hub and are unable to pitch, which leads to aerodynamic stall above certain windspeeds. In other more sophisticated designs, they are bolted to the pitch mechanism, which adjusts their angle of attack according to the wind speed to control their rotational speed. The pitch mechanism is itself bolted to the hub. The hub is fixed to the rotor shaft which drives the generator directly or through a gearbox.

Blade Count

The 98 meter diameter, two-bladed NASA/DOE Mod-5B wind turbine was the largest
operating wind turbine in the world in the early 1990s.

The number of blades is selected for aerodynamic efficiency, component costs, and system reliability. Noise emissions are affected by the location of the blades upwind or downwind of the tower and the speed of the rotor. Given that the noise emissions from the blades' trailing edges and tips vary by the 5th power of blade speed, a small increase in tip speed can make a large difference.

Wind turbines developed over the last 50 years have almost universally used either two or three blades. However, there are patents that present designs with additional blades, such as Chan

Shin's Multi-unit rotor blade system integrated wind turbine. Aerodynamic efficiency increases with number of blades but with diminishing return. Increasing the number of blades from one to two yields a six percent increase in aerodynamic efficiency, whereas increasing the blade count from two to three yields only an additional three percent in efficiency. Further increasing the blade count yields minimal improvements in aerodynamic efficiency and sacrifices too much in blade stiffness as the blades become thinner.

The NASA test of a one-bladed wind turbine rotor configuration at Plum Brook Station near Sandusky, Ohio.

Theoretically, an infinite number of blades of zero width is the most efficient, operating at a high value of the tip speed ratio. But other considerations lead to a compromise of only a few blades.

Component costs that are affected by blade count are primarily for materials and manufacturing of the turbine rotor and drive train. Generally, the lower the number of blades, the lower the material and manufacturing costs will be. In addition, the lower the number of blades, the higher the rotational speed can be. This is because blade stiffness requirements to avoid interference with the tower limit how thin the blades can be manufactured, but only for upwind machines; deflection of blades in a downwind machine results in increased tower clearance. Fewer blades with higher rotational speeds reduce peak torques in the drive train, resulting in lower gearbox and generator costs.

System reliability is affected by blade count primarily through the dynamic loading of the rotor into the drive train and tower systems. While aligning the wind turbine to changes in wind direction (yawing), each blade experiences a cyclic load at its root end depending on blade position. This is true of one, two, three blades or more. However, these cyclic loads when combined together at the drive train shaft are symmetrically balanced for three blades, yielding smoother operation during turbine yaw. Turbines with one or two blades can use a pivoting teetered hub to also nearly eliminate the cyclic loads into the drive shaft and system during yawing. A Chinese 3.6 MW two-blade is being tested in Denmark. Mingyang won a bid for 87 MW (29 * 3 MW) two-bladed offshore wind turbines near Zhuhai in 2013.

Finally, aesthetics can be considered a factor in that some people find that the three-bladed rotor is more pleasing to look at than a one- or two-bladed rotor.

Blade Materials

Several modern wind turbines use rotor blades with carbon-fibre girders to reduce weight.

In general, ideal materials should meet the following criteria:

- wide availability and easy processing to reduce cost and maintenance

- low weight or density to reduce gravitational forces

- high strength to withstand strong loading of wind and gravitational force of the blade itself

- high fatigue resistance to withstand cyclic loading

- high stiffness to ensure stability of the optimal shape and orientation of the blade and clearance with the tower

- high fracture toughness

- the ability to withstand environmental impacts such as lightning strikes, humidity, and temperature

This narrows down the list of acceptable materials. Metals would be undesirable because of their vulnerability to fatigue. Ceramics have low fracture toughness, which could result in early blade failure. Traditional polymers are not stiff enough to be useful, and wood has problems with repeatability, especially considering the length of the blade. That leaves fiber-reinforced composites, which have high strength and stiffness and low density, as a very attractive class of materials for the design of wind turbines.

Wood and canvas sails were used on early windmills due to their low price, availability, and ease of manufacture. Smaller blades can be made from light metals such as aluminium. These materials, however, require frequent maintenance. Wood and canvas construction limits the airfoil shape to a flat plate, which has a relatively high ratio of drag to force captured (low aerodynamic efficiency) compared to solid airfoils. Construction of solid airfoil designs requires inflexible materials such as metals or composites. Some blades also have incorporated lightning conductors.

New wind turbine designs push power generation from the single megawatt range to upwards of 10 megawatts using larger and larger blades. A larger area effectively increases the tip-speed ratio of a turbine at a given wind speed, thus increasing its energy extraction. Computer-aided engineering software such as HyperSizer (originally developed for spacecraft design) can be used to improve blade design.

As of 2015 the rotor diameters of onshore wind turbine blades are as large as 130 meters, while the diameter of offshore turbines reach 170 meters. In 2001, an estimated 50 million kilograms of fibreglass laminate were used in wind turbine blades.

An important goal of larger blade systems is to control blade weight. Since blade mass scales as the cube of the turbine radius, loading due to gravity constrains systems with larger blades. Gravitational loads include axial and tensile/ compressive loads (top/bottom of rotation) as well as bending (lateral positions). The magnitude of these loads fluctuates cyclically and the edgewise moments are reversed every 180° of rotation. Typical rotor speeds and design life are ~10rpm and 20 years, respectively, with the number of lifetime revolutions on the order of 10^8. Considering wind, it is expected that turbine blades go through ~10^9 loading cycles. Wind is another source of rotor blade loading. Lift causes bending in the flapwise direction (out of rotor plane) while air flow around the blade cause edgewise bending (in the rotor plane). Flapwise bending involves tension on the pressure (upwind) side and compression on the suction (downwind) side. Edgewise bending involves tension on the leading edge and compression on the trailing edge.

Wind loads are cyclical because of natural variability in wind speed and wind shear (higher speeds at top of rotation).

Failure in ultimate loading of wind-turbine rotor blades exposed to wind and gravity loading is a failure mode that needs to be considered when the rotor blades are designed. The wind speed that causes bending of the rotor blades exhibits a natural variability, and so does the stress response in the rotor blades. Also, the resistance of the rotor blades, in terms of their tensile strengths, exhibits a natural variability.

In light of these failure modes and increasingly larger blade systems, there has been continuous effort toward developing cost-effective materials with higher strength-to-mass ratios. In order to extend the current 20 year lifetime of blades and enable larger area blades to be cost-effective, the design and materials need to be optimized for stiffness, strength, and fatigue resistance.

The majority of current commercialized wind turbine blades are made from fiber-reinforced polymers (FRPs), which are composites consisting of a polymer matrix and fibers. The long fibers provide longitudinal stiffness and strength, and the matrix provides fracture toughness, delamination strength, out-of-plane strength, and stiffness. Material indices based on maximizing power efficiency, and having high fracture toughness, fatigue resistance, and thermal stability, have been shown to be highest for glass and carbon fiber reinforced plastics (GFRPs and CFRPs).

Fiberglass-reinforced epoxy blades of Siemens SWT-2.3-101 wind turbines. The blade size of 49 meters is in comparison to a substation behind them at Wolfe Island Wind Farm.

Manufacturing blades in the 40 to 50 metre range involves proven fibreglass composite fabrication techniques. Manufactures such as Nordex SE and GE Wind use an infusion process. Other manufacturers use variations on this technique, some including carbon and wood with fibreglass in an epoxy matrix. Other options include preimpregnated ("prepreg") fibreglass and vacuum-assisted resin transfer molding. Each of these options use a glass-fibre reinforced polymer composite constructed with differing complexity. Perhaps the largest issue with more simplistic, open-mould, wet systems are the emissions associated with the volatile organics released. Preimpregnated materials and resin infusion techniques avoid the release of volatiles by containing all VOCs. However, these contained processes have their own challenges, namely the production of thick laminates necessary for structural components becomes more difficult. As the preform resin permeability dictates the maximum laminate thickness, bleeding is required to eliminate voids and ensure proper resin distribution. One solution to resin distribution a partially preimpregnated fibreglass. During evacuation, the dry fabric provides a path for airflow and, once heat and pressure are applied, resin may flow into the dry region resulting in a thoroughly impregnated laminate structure.

Epoxy-based composites have environmental, production, and cost advantages over other resin systems. Epoxies also allow shorter cure cycles, increased durability, and improved surface finish. Prepreg operations further reduce processing time over wet lay-up systems. As turbine blades pass 60 metres, infusion techniques become more prevalent; the traditional resin transfer moulding injection time is too long as compared to the resin set-up time, limiting laminate thickness. Injection forces resin through a thicker ply stack, thus depositing the resin where in the laminate structure before gelation occurs. Specialized epoxy resins have been developed to customize lifetimes and viscosity.

Carbon fibre-reinforced load-bearing spars can reduce weight and increase stiffness. Using carbon fibres in 60 metre turbine blades is estimated to reduce total blade mass by 38% and decrease cost by 14% compared to 100% fibreglass. Carbon fibres have the added benefit of reducing the thickness of fiberglass laminate sections, further addressing the problems associated with resin wetting of thick lay-up sections. Wind turbines may also benefit from the general trend of increasing use and decreasing cost of carbon fibre materials.

Although glass and carbon fibers have many optimal qualities for turbine blade performance, there are several downsides to these current fillers, including the fact that high filler fraction (10-70 wt%) causes increased density as well as microscopic defects and voids that often lead to premature failure.

Recent developments include interest in using carbon nanotubes (CNT's) to reinforce polymer-based nanocomposites. CNT's can be grown or deposited on the fibers, or added into polymer resins as a matrix for FRP structures. Using nanoscale CNT's as filler instead of traditional microscale filler (such as glass or carbon fibers) results in CNT/polymer nanocomposites, for which the properties can be changed significantly at very low filler contents (typically < 5 wt%). They have very low density, and improve the elastic modulus, strength, and fracture toughness of the polymer matrix. The addition of CNT's to the matrix also reduces the propagation of interlaminar cracks which can be a problem in traditional FRP's.

Further improvement is possible through the use of carbon nanofibers (CNFs) in the blade coatings. A major problem in desert environments is erosion of the leading edges of blades by wind carrying sand, which increases roughness and decreases aerodynamic performance. The particle erosion resistance of fiber-reinforced polymers is poor when compared to metallic materials and elastomers, and needs to be improved. It has been shown that the replacement of glass fiber with CNF on the composite surface greatly improves erosion resistance. CNF's have also been shown to provide good electrical conductivity (important for lightning strikes), high damping ratio, and good impact-friction resistance. These properties make CNF-based nanopaper a prospective coating for wind turbine blades.

Blade Recycling

The Global Wind Energy Council (GWEC) predicts that wind energy will supply 15.7% of the world's total energy needs by the year 2020, and 28.5% by the year 2030. This dramatic increase in global wind energy generation will require installation of a newer and larger fleet of more efficient wind turbines and the consequent decommissioning of aging ones. Based on a study carried out by the European Wind Energy Association, in the year 2010 alone, between 110 and 140 kilotons of composites were consumed by the wind turbine industry for manufacturing blades. The majority of the blade material will eventually end up as waste, and in order to accommodate this level of composite waste, the only option is recycling. Typically, glass-fibre-reinforced-polymers (GFRPs) compose of around 70% of the laminate material in the blade. GFRPs hinder incineration and are not combustible. Therefore, conventional recycling methods need to be modified. Currently, depending on whether individual fibres can be recovered, there exists a few general methods for recycling GFRPs in wind turbine blades:

- Mechanical Recycling: This method doesn't recover individual fibres. Initial processes involve shredding, crushing, and/or milling. The crushed pieces are then separated into fibre-rich and resin-rich fractions. These fractions are ultimately incorporated into new composites either as fillers or reinforcements.

- Chemical Processing/Pyrolysis: Thermal decomposition of the composites is used to recover the individual fibres. For pyrolysis, the material is heated up to 500 °C in an environment without oxygen, thus causing it to break down into lower weight organic substances and/or gaseous products. The glass fibres will generally loose 50% of their initial strength and can now be downcycled for fibre reinforcement applications in paints or concrete. Research has shown that this end of life option is able to recover up to approximately 19 MJ/kg. However, this method has a relatively high cost and requires similar mechanical pre-processing. In addition, it has not yet been modified to satisfy the future need of large scale wind turbine blade recycling.

Tower

Two main types of towers exist: floating towers and land-based towers, which are usually more common.

Tower Height

Wind velocities increase at higher altitudes due to surface aerodynamic drag (by land or water surfaces) and the viscosity of the air. The variation in velocity with altitude, called wind shear, is most dramatic near the surface. Typically, the variation follows the wind profile power law, which predicts that wind speed rises proportionally to the seventh root of altitude. Doubling the altitude of a turbine, then, increases the expected wind speeds by 10% and the expected power by 34%. To avoid buckling, doubling the tower height generally requires doubling the diameter of the tower as well, increasing the amount of material by a factor of at least four.

At night time, or when the atmosphere becomes stable, wind speed close to the ground usually subsides whereas at turbine hub altitude it does not decrease that much or may even increase. As a result, the wind speed is higher and a turbine will produce more power than expected from the 1/7 power law: doubling the altitude may increase wind speed by 20% to 60%. A stable atmosphere is caused by radiative cooling of the surface and is common in a temperate climate: it usually occurs when there is a (partly) clear sky at night. When the (high altitude) wind is strong (a 10-meter wind speed higher than approximately 6 to 7 m/s) the stable atmosphere is disrupted because of friction turbulence and the atmosphere will turn neutral. A daytime atmosphere is either neutral (no net radiation; usually with strong winds and heavy clouding) or unstable (rising air because of ground heating—by the sun). Here again the 1/7 power law applies or is at least a good approximation of the wind profile. Indiana had been rated as having a wind capacity of 30,000 MW, but by raising the expected turbine height from 50 m to 70 m, the wind capacity estimate was raised to 40,000 MW, and could be double that at 100 m.

For HAWTs, tower heights approximately two to three times the blade length have been found to balance material costs of the tower against better utilisation of the more expensive active components.

Sections of a wind turbine tower, transported in a bulk carrier ship.

Road size restrictions makes transportation of towers with a diameter of more than 4.3 m difficult. Swedish analyses show that it is important to have the bottom wing tip at least 30 m above the tree tops, but a taller tower requires a larger tower diameter. A 3 MW turbine may increase output from 5,000 MWh to 7,700 MWh per year by going from 80 to 125 meter tower height. A tower profile made of connected shells rather than cylinders can have a larger diameter and still be transportable. A 100 m prototype tower with TC bolted 18 mm 'plank' shells has been erected at the wind turbine test center Høvsøre in Denmark and certified by Det Norske Veritas, with a Siemens nacelle. Shell elements can be shipped in standard 12 m shipping containers, and 2½ towers per week are produced this way.

As of 2003, typical modern wind turbine installations use towers about 210 ft (65 m) high. Height is typically limited by the availability of cranes. This has led to a variety of proposals for "partially self-erecting wind turbines" that, for a given available crane, allow taller towers that put a turbine in stronger and steadier winds, and "self-erecting wind turbines" that can be installed without cranes.

Tower Materials

Currently, the majority of wind turbines are supported by conical tubular steel towers. These towers represent 30% – 65% of the turbine weight and therefore account for a large percentage of the turbine transportation costs. The use of lighter materials in the tower could greatly reduce the overall transport and construction cost of wind turbines, however the stability must be maintained. Higher grade S500 steel costs 20%-25% more than S335 steel (standard structural steel), but it requires 30% less material because of its improved strength. Therefore, replacing wind turbine towers with S500 steel would result in a net savings in both weight and cost.

Another disadvantage of conical steel towers is that constructing towers that meet the requirements of wind turbines taller than 90 meters proves challenging. High performance concrete shows potential to increase tower height and increase the lifetime of the towers. A hybrid of prestressed concrete and steel has shown improved performance over standard tubular steel at tower heights of 120 meters. Concrete also gives the benefit of allowing for small precast sections to be assembled on site, avoiding the challenges steel faces during transportation. One downside of concrete towers is the higher CO_2 emissions during concrete production as compared to steel. However, the overall environmental benefit should be higher if concrete towers can double the wind turbine lifetime.

Wood is being investigated as a material for wind turbine towers, and a 100 metre tall tower supporting a 1.5 MW turbine has been erected in Germany. The wood tower shares the same transportation benefits of the segmented steel shell tower, but without the steel resource consumption.

Connection to the Electric Grid

All grid-connected wind turbines, from the first one in 1939 until the development of variable-speed grid-connected wind turbines in the 1970s, were fixed-speed wind turbines. As recently as 2003, nearly all grid-connected wind turbines operated at exactly constant speed (synchronous generators) or within a few percent of constant speed (induction generators). As of 2011, many operational wind turbines used fixed speed induction generators (FSIG). As of 2011, most new grid-connected wind turbines are variable speed wind turbines—they are in some variable speed configuration.

Early wind turbine control systems were designed for peak power extraction, also called maximum power point tracking—they attempt to pull the maximum possible electrical power from a given wind turbine under the current wind conditions. More recent wind turbine control systems deliberately pull less electrical power than they possibly could in most circumstances, in order to provide other benefits, which include:

- spinning reserves to quickly produce more power when needed—such as when some other generator suddenly drops from the grid—up to the max power supported by the current wind conditions.

- Variable-speed wind turbines can (very briefly) produce more power than the current wind conditions can support, by storing some wind energy as kinetic energy (accelerating during brief gusts of faster wind) and later converting that kinetic energy to electric energy (decelerating, either when more power is needed elsewhere, or during short lulls in the wind, or both).

- damping (electrical) subsynchronous resonances in the grid

- damping (mechanical) resonances in the tower

The generator in a wind turbine produces alternating current (AC) electricity. Some turbines drive an AC/AC converter—which converts the AC to direct current (DC) with a rectifier and then back to AC with an inverter—in order to match the frequency and phase of the grid. However, the most common method in large modern turbines is to instead use a doubly fed induction generator directly connected to the electricity grid.

A useful technique to connect a permanent magnet synchronous generator to the grid is by using a back-to-back converter. Also, we can have control schemes so as to achieve unity power factor in the connection to the grid. In that way the wind turbine will not consume reactive power, which is the most common problem with wind turbines that use induction machines. This leads to a more stable power system. Moreover, with different control schemes a wind turbine with a permanent magnet synchronous generator can provide or consume reactive power. So, it can work as a dynamic capacitor/inductor bank so as to help with the power systems' stability.

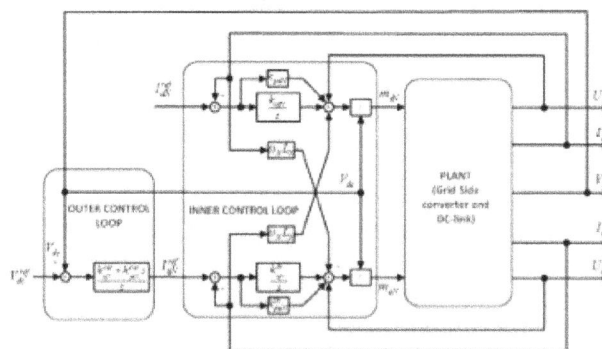

Grid Side Controller Design

Below we show the control scheme so as to achieve unity power factor :

Reactive power regulation consists of one PI controller in order to achieve operation with unity power factor (i.e. $Q_{grid} = 0$). It is obvious that I_{dN} has to be regulated to reach zero at steady-state ($I_{dNref} = 0$).

We can see the complete system of the grid side converter and the cascaded PI controller loops in the figure.

Foundations

Wind turbine foundations

Wind turbines, by their nature, are very tall slender structures, this can cause a number of issues when the structural design of the foundations are considered.

The foundations for a conventional engineering structure are designed mainly to transfer the vertical load (dead weight) to the ground, this generally allows for a comparatively unsophisticated arrangement to be used. However, in the case of wind turbines, due to the high wind and environmental loads experienced there is a significant horizontal dynamic load that needs to be appropriately restrained.

This loading regime causes large moment loads to be applied to the foundations of a wind turbine. As a result, considerable attention needs to be given when designing the footings to ensure that the turbines are sufficiently restrained to operate efficiently. In the current Det Norske Veritas (DNV) guidelines for the design of wind turbines the angular deflection of the foundations are limited to 0.5°. DNV guidelines regarding earthquakes suggest that horizontal loads are larger than vertical loads for offshore wind turbines, while guidelines for tsunamis only suggest designing for maximum sea waves. In contrast, IEC suggests considering tsunami loads.

Scale model tests using a 50-g centrifuge are being performed at the Technical University of Denmark to test monopile foundations for offshore wind turbines at 30 to 50-m water depth.

Costs

Liftra *Blade Dragon* installing a single blade on wind turbine hub.

The modern wind turbine is a complex and integrated system. Structural elements comprise the majority of the weight and cost. All parts of the structure must be inexpensive, lightweight, durable, and manufacturable, under variable loading and environmental conditions. Turbine systems that have fewer failures, require less maintenance, are lighter and last longer will lead to reducing the cost of wind energy.

One way to achieve this is to implement well-documented, validated analysis codes, according to a 2011 report from a coalition of researchers from universities, industry, and government, supported by the Atkinson Center for a Sustainable Future.

The major parts of a modern turbine may cost (percentage of total): tower 22%, blades 18%, gearbox 14%, generator 8%.

Efficiency and Wind Speed

The efficiency of a wind turbine is maximum at its design wind velocity, and efficiency decreases with the fluctuations in wind. The lowest velocity at which the turbine develops its full power is known as rated wind velocity. Below some minimum wind velocity, no useful power output can be produced from wind turbine. There are limits on both the minimum (2–5 m/s) and maximum (25–30 m/s) wind velocity for the efficient operation of wind turbines.

Conservation of mass requires that the amount of air entering and exiting a turbine must be equal. Accordingly, Betz's law gives the maximal achievable extraction of wind power by a wind turbine as 16/27 (59.3%) of the total kinetic energy of the air flowing through the turbine.

The maximum theoretical power output of a wind machine is thus 0.59 times the kinetic energy of the air passing through the effective disk area of the machine. If the effective area of the disk is A, and the wind velocity v, the maximum theoretical power output P is:

$$P = 0.59 \frac{1}{2} \rho v^3 A$$

where ρ is air density

As wind is free (no fuel cost), wind-to-rotor efficiency (including rotor blade friction and drag) is one of many aspects impacting the final price of wind power. Further inefficiencies, such as gearbox losses, generator and converter losses, reduce the power delivered by a wind turbine. To protect components from undue wear, extracted power is held constant above the rated operating speed as theoretical power increases at the cube of wind speed, further reducing theoretical efficiency. In 2001, commercial utility-connected turbines deliver 75% to 80% of the Betz limit of power extractable from the wind, at rated operating speed.

All power plants have some consumption when they produce power, and some standby consumption when they are turned on without producing power. For a modern 3 MW wind turbine, the consumption may be 6-58 kW depending on circumstances.

Design Specification

The design specification for a wind-turbine will contain a power curve and guaranteed availability. With the data from the wind resource assessment it is possible to calculate commercial viability. The typical operating temperature range is −20 to 40 °C (−4 to 104 °F). In areas with extreme climate (like Inner Mongolia or Rajasthan) specific cold and hot weather versions are required.

Wind turbines can be designed and validated according to IEC 61400 standards.

Low Temperature

Utility-scale wind turbine generators have minimum temperature operating limits which apply in areas that experience temperatures below −20 °C. Wind turbines must be protected from ice accumulation. It can make anemometer readings inaccurate and which, in certain turbine control designs, can cause high structure loads and damage. Some turbine manufacturers offer low-temperature packages at a few percent extra cost, which include internal heaters, different lubricants, and different alloys for structural elements. If the low-temperature interval is combined with a low-wind condition, the wind turbine will require an external supply of power, equivalent to a few percent of its rated power, for internal heating. For example, the St. Leon, Manitoba project has a total rating of 99 MW and is estimated to need up to 3 MW (around 3% of capacity) of station service power a few days a year for temperatures down to −30 °C. This factor affects the economics of wind turbine operation in cold climates.

Thermoelectric Generator

A thermoelectric generator (TEG), also called a Seebeck generator, is a solid state device that converts heat flux (temperature differences) directly into electrical energy through a phenomenon called the Seebeck effect (a form of thermoelectric effect). Thermoelectric generators function like heat engines, but are less bulky and have no moving parts. However, TEGs are typically more expensive and less efficient.

Thermoelectric generators could be used in power plants in order to convert waste heat into additional electrical power and in automobiles as automotive thermoelectric generators (ATGs) to increase fuel efficiency. Another application is radioisotope thermoelectric generators which are used in space probes, which has the same mechanism but use radioisotopes to generate the required heat difference.

History

In 1821, Thomas Johann Seebeck discovered that a thermal gradient formed between two dissimilar conductors can produce electricity. At the heart of the thermoelectric effect is the fact that a temperature gradient in a conducting material results in heat flow; this results in the diffusion of charge carriers. The flow of charge carriers between the hot and cold regions in turn creates a voltage difference. In 1834, Jean Charles Athanase Peltier discovered the reverse effect, that running an electric current through the junction of two dissimilar conductors could, depending on the direction of the current, cause it to act as a heater or cooler.

Construction

Thermoelectric power generators consist of three major components: thermoelectric materials, thermoelectric modules and thermoelectric systems that interface with the heat source.

Thermoelectric Materials

Thermoelectric materials generate power directly from heat by converting temperature differences into electric voltage. These materials must have both high electrical conductivity (σ) and low thermal conductivity (κ) to be good thermoelectric materials. Having low thermal conductivity ensures that when one side is made hot, the other side stays cold, which helps to generate a large voltage while in a temperature gradient. The measure of the magnitude of electrons flow in response to a temperature difference across that material is given by the Seebeck coefficient (S). The efficiency of a given material to produce a thermoelectric power is governed by its "figure of merit" $zT = S^2\sigma T/\kappa$.

For many years, the main three semiconductors known to have both low thermal conductivity and high power factor were bismuth telluride (Bi_2Te_3), lead telluride (PbTe), and silicon germanium (SiGe). These materials have very rare elements which make them very expensive compounds.

Today, the thermal conductivity of semiconductors can be lowered without affecting their high electrical properties using nanotechnology. This can be achieved by creating nanoscale features such as particles, wires or interfaces in bulk semiconductor materials. However, the manufacturing processes of nano-materials is still challenging.

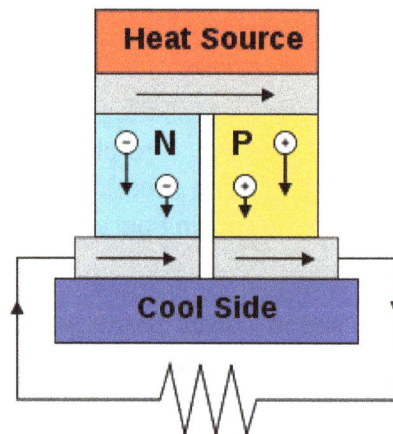

A thermoelectric circuit composed of materials of different Seebeck coefficient (p-doped and n-doped semiconductors), configured as a thermoelectric generator.

Thermoelectric Module

A thermoelectric module is a circuit containing thermoelectric materials that generate electricity from heat directly. A thermoelectric module consists of two dissimilar thermoelectric materials joining in their ends: an n-type (negatively charged); and a p-type (positively charged) semiconductors. A direct electric current will flow in the circuit when there is a temperature difference between the two materials. Generally, the current magnitude has a proportional relationship with the temperature difference. (i.e., the more the temperature difference, the higher the current.)

In application, thermoelectric modules in power generation work in very tough mechanical and thermal conditions. Because they operate in very high temperature gradient, the modules are subject to large thermally induced stresses and strains for long periods of time. They also are subject to mechanical fatigue caused by large number of thermal cycles.

Thus, the junctions and materials must be selected so that they survive these tough mechanical and thermal conditions. Also, the module must be designed such that the two thermoelectric materials are thermally in parallel, but electrically in series. The efficiency of thermoelectric modules are greatly affected by its geometrical design.

Thermoelectric System

Using thermoelectric modules, a thermoelectric system generates power by taking in heat from a source such as a hot exhaust flue. In order to do that, the system needs a large temperature gradient, which is not easy in real-world applications. The cold side must be cooled by air or water. Heat exchangers are used on both sides of the modules to supply this heating and cooling.

There are many challenges in designing a reliable TEG system that operates at high temperatures. Achieving high efficiency in the system requires extensive engineering design in order to balance between the heat flow through the modules and maximizing the temperature gradient across them. To do this, designing heat exchanger technologies in the system is one of the most important aspects of TEG engineering. In addition, the system requires to minimize the thermal losses due to the interfaces between materials at several places. Another challenging constraint is avoiding large pressure drops between the heating and cooling sources.

After the DC power from the TE modules passes through an inverter, the TEG produces AC power, which in turn, requires an integrated power electronics system to deliver it to the customer.

Materials for TEG

Only a few known materials to date are identified as thermoelectric materials. Most thermoelectric materials today have a ZT, the figure of merit, value of around unity, such as in Bismuth Telluride (Bi_2Te_3) at room temperature and lead telluride (PbTe) at 500-700K. However, in order to be competitive with other power generation systems, TEG materials should have zT of 2-3 range. Most research in thermoelectric materials has focused on increasing the Seebeck coefficient (S) and reducing the thermal conductivity, especially by manipulating the nanostructure of the thermoelectric materials. Because the thermal and electrical conductivity correlate with the charge carriers, new means must be introduced in order to conciliate the contradiction between high electrical conductivity and low thermal conductivity as indicated.

When selecting materials for thermoelectric generation, a number of other factors need to be considered. During operation, ideally the thermoelectric generator has a large temperature gradient across it. Thermal expansion will then introduce stress in the device which may cause fracture of the thermoelectric legs, or separation from the coupling material. The mechanical properties of the materials must be considered and the coefficient of thermal expansion of the n and p-type material must be matched reasonably well. In segmented thermoelectric generators, the material's compatibility must also be considered. A material's compatibility factor is defined as $s = \left(\dfrac{\sqrt{1-zT}-1}{ST} \right)$.

When the compatibility factor from one segment to the next differs by more than a factor of about two, the device will not operate efficiently. The material parameters determining s (as well as zT) are temperature dependent, so the compatibility factor may change from the hot side to the cold side of the device, even in one segment. This behavior is referred to as self-compatibility and may become important in devices design for low temperature operation.

In general, thermoelectric materials can be categorized into conventional and new materials:

Conventional Materials

There are many TEG materials that are employed in commercial applications today. These materials can be divided into three groups based on the temperature range of operation:

1. Low temperature materials (up to around 450K): Alloys based on Bismuth (Bi) in combinations with Antimony (Sb), Tellurium (Te) or Selenium (Se).

2. Intermediate temperature (up to 850K): such as materials based on alloys of Lead (Pb)

3. Highest temperatures material (up to 1300K): materials fabricated from silicon germanium (SiGe) alloys.

Although these materials still remain the cornerstone for commercial and practical applications in thermoelectric power generation, significant advances have been made in synthesizing new materials and fabricating material structures with improved thermoelectric performance. Recent research have focused on improving the material's figure-of-merit (zT), and hence the conversion efficiency, by reducing the lattice thermal conductivity.

New Materials

Researchers are trying to develop new thermoelectric materials for power generation by improving the figure-of-merit zT. One example of these materials is the semiconductor compound β-Zn_4Sb_3, which possesses an exceptionally low thermal conductivity and exhibits a maximum zT of 1.3 at a temperature of 670K. This material is also relatively inexpensive and stable up to this temperature in a vacuum, and can be a good alternative in the temperature range between materials based on Bi_2Te_3 and PbTe.

Beside improving the figure-of-merit, there is increasing focus to develop new materials by increasing the electrical power output, decreasing cost and developing environmentally friendly materials. For example, when the fuel cost is low or almost free, such as in waste heat recovery, then the cost per watt is only determined by the power per unit area and the operating period. As a result, it has initiated a search for materials with high power output rather than conversion efficiency. For example, the rare earth compounds $YbAl_3$ has a low figure-of-merit, but it has a power output of at least double that of any other material, and can operate over the temperature range of a waste heat source.

Novel Processing

In order to increase the figure of merit (zT), a material's thermal conductivity should be minimized while its electrical conductivity and Seebeck coefficient is maximized. In most cases, methods to increase or decrease one property result in the same effect on other properties due to their interdependence. A novel processing technique exploits the scattering of different phonon frequencies to selectively reduce lattice thermal conductivity without the typical negative effects on electrical conductivity from the simultaneous increased scattering of electrons. In a bismuth antimony tellurium ternary system, liquid-phase sintering is used to produce low-energy semicoherent grain boundaries, which do not have a significant scattering effect on electrons. The breakthrough is then applying a pressure to the liquid in the sintering process, which creates a transient flow of the Te rich liquid and facilitates the formation of dislocations that greatly reduce the lattice conductivity. The ability to selectively decrease the lattice conductivity results in reported zT values of $1.86 \pm .15$ which are a significant improvement over current commercial thermoelectric generators which have typical figures of merit closer to .3-.6. These improvements highlight the fact that in addition to development of novel materials for thermoelectric applications, using different processing techniques to design microstructure is a viable and worthwhile effort. In fact, it often makes sense to work to optimize both composition and microstructure.

Efficiency

The typical efficiency of TEGs is around 5–8%. Older devices used bimetallic junctions and were bulky. More recent devices use highly doped semiconductors made from bismuth telluride (Bi_2Te_3), lead telluride (PbTe), calcium manganese oxide ($Ca_2Mn_3O_8$), or combinations thereof, depending on temperature. These are solid-state devices and unlike dynamos have no moving parts, with the occasional exception of a fan or pump.

Uses

Thermoelectric generators have a variety of applications. Frequently, thermoelectric generators are used for low power remote applications or where bulkier but more efficient heat engines such as Stirling engines would not be possible. Unlike heat engines, the solid state electrical components typically used to perform thermal to electric energy conversion have no moving parts. The thermal to electric energy conversion can be performed using components that require no maintenance, have inherently high reliability, and can be used to construct generators with long service-free lifetimes. This makes thermoelectric generators well suited for equipment with low to modest power needs in remote uninhabited or inaccessible locations such as mountaintops, the vacuum of space, or the deep ocean.

- Common application is the use of thermoelectric generators on gas pipelines. For example, for cathodic protection, radio communication, and other telemetry. On gas pipelines for power consumption of up to 5 kW thermal generators are preferable to other power sources. The manufacturers of generators for gas pipelines are Gentherm Global Power Technologies (Formerly Global Thermoelectric), (Calgary, Canada) and TELGEN (Russia).

- Thermoelectric Generators are primarily used as remote and off-grid power generators for unmanned sites. They are the most reliable power generator in such situations as they do not have moving parts (thus virtually maintenance free), work day and night, perform under all weather conditions, and can work without battery backup. Although Solar Photovoltaic systems are also implemented in remote sites, Solar PV may not be a suitable solution where solar radiation is low, i.e. areas at higher latitudes with snow or no sunshine, areas with lots of cloud or tree canopy cover, dusty deserts, forests, etc.

- Gentherm Global Power Technologies (GPT) formerly known as Global Thermoelectric (Canada) has Hybrid Solar-TEG solutions where the Thermoelectric Generator backs up the Solar-PV, such that if the Solar panel is down and the backup battery backup goes into deep discharge then a sensor starts the TEG as a backup power source until the Solar is up again. The TEG heat can be produced by a low pressure flame fueled by Propane or Natural Gas.

- Many space probes, including the Mars *Curiosity* rover, generate electricity using a radio-isotope thermoelectric generator whose heat source is a radioactive element.

- Cars and other automobiles produce waste heat (in the exhaust and in the cooling agents). Harvesting that heat energy, using a thermoelectric generator, can increase the fuel efficiency of the car.

- In addition to automobiles, waste heat is also generated in many other places, such as in industrial processes and in heating (wood stoves, outdoor boilers, cooking, oil and gas fields, pipelines, and remote communication towers).

- Microprocessors generate waste heat. Researchers have considered whether some of that energy could be recycled.

- Solar cells use only the high frequency part of the radiation, while the low frequency heat energy is wasted. Several patents about the use of thermoelectric devices in tandem with solar cells have been filed. The idea is to increase the efficiency of the combined solar/thermoelectric system to convert the solar radiation into useful electricity.

- The Maritime Applied Physics Corporation in Baltimore, Maryland is developing a thermoelectric generator to produce electric power on the deep-ocean offshore seabed using the temperature difference between cold seawater and hot fluids released by hydrothermal vents, hot seeps, or from drilled geothermal wells. A high reliability source of seafloor electric power is needed for ocean observatories and sensors used in the geological, environmental, and ocean sciences, by seafloor mineral and energy resource developers, and by the military.

- Ann Makosinski from British Columbia, Canada has developed several devices using Peltier tiles to harvest heat (from a human hand, the forehead, and hot beverage) that claims to generate enough electricity to power an LED light or charge a mobile device, although the inventor admits that the brightness of the LED light is not competitive with those on the market.

Practical Limitations

Besides low efficiency and relatively high cost, practical problems exist in using thermoelectric devices in certain types of applications resulting from a relatively high electrical output resistance, which increases self-heating, and a relatively low thermal conductivity, which makes them unsuitable for applications where heat removal is critical, as with heat removal from an electrical device such as microprocessors.

- High generator output resistance: In order to get voltage output levels in the range required by digital electrical devices, a common approach is to place many thermoelectric elements in series within a generator module. The element's voltages add, but so do their individual output resistance. The maximum power transfer theorem dictates that maximum power is delivered to a load when the source and load resistances are identically matched. For low impedance loads near zero ohms, as the generator resistance rises the power delivered to the load decreases. To lower the output resistance, some commercial devices place more individual elements in parallel and fewer in series and employ a boost regulator to raise the voltage to the voltage needed by the load.

- Low thermal conductivity: Because a very high thermal conductivity is required to transport thermal energy away from a heat source such as a digital microprocessor, the low thermal conductivity of thermoelectric generators makes them unsuitable to recover the heat.

- Cold-side heat removal with air: In air-cooled thermoelectric applications, such as when harvesting thermal energy from a motor vehicle's crankcase, the large amount of thermal energy that must be dissipated into ambient air presents a significant challenge. As a thermoelectric generator's cool side temperature rises, the device's differential working temperature decreases. As the temperature rises, the device's electrical resistance increases causing greater parasitic generator self-heating. In motor vehicle applications a supplementary radiator is sometimes used for improved heat removal, though the use of an electric water pump to circulate a coolant adds an additional parasitic loss to total generator output power. Water cooling the thermoelectric generator's cold side, as when generating thermoelectric power from the hot crank case of an inboard boat motor, would not suffer from this disadvantage. Water is a far easier coolant to use effectively in contrast to air.

Future Market

While TEG technology has been used in military and aerospace applications for decades, new TE materials and systems are being developed to generate power using low or high temperatures waste heat, and that could provide a significant opportunity in the near future . These systems can also be scalable to any size and have lower operation and maintenance cost.

In general, investing in TEG technology is increasing rapidly. The global market for thermoelectric generators is estimated to be US$320 million in 2015. A recent study estimated that TEG is expected to reach $720 million in 2021 with a growth rate of 14.5%. Today, North America capture 66% of the market share and it will continue to be the biggest market in the near future. However, Asia-Pacific and European countries are projected to grow at relatively higher rates. A study found that the Asia-Pacific market would grow at a Compound Annual Growth Rate (CAGR) of 18.3% in the period from 2015 to 2020 due to the high demand of thermoelectric generators by the automotive industries to increase overall fuel efficiency, as well as the growing industrialization in the region.

Low power TEG or "Sub-watt" (i.e. generating up to 1 Watt peak) market is a growing part of the TEG market, capitalizing on latest technologies. Main applications are sensors, low power applications and more globally Internet of things applications. A specialized market research company indicated that 100 000 units have been shipped in 2014 and expects 9 million units per year by 2020.

Automotive Thermoelectric Generator

An automotive thermoelectric generator (ATEG) is a device that converts some of the waste heat of an internal combustion engine (IC) into electricity using the Seebeck Effect. A typical ATEG consists of four main elements: A hot-side heat exchanger, a cold-side heat exchanger, thermoelectric materials, and a compression assembly system. ATEGs can convert waste heat from an engine's coolant or exhaust into electricity. By reclaiming this otherwise lost energy, ATEGs decrease fuel consumed by the electric generator load on the engine. However, the cost of the unit and the extra fuel consumed due to its weight must be also considered.

Operation Principles

In ATEGs, thermoelectric materials are packed between the hot-side and the cold-side heat exchangers. The thermoelectric materials are made up of p-type and n-type semiconductors, while the heat exchangers are metal plates with high thermal conductivity.

The temperature difference between the two surfaces of the thermoelectric module(s) generates electricity using the Seebeck Effect. When hot exhaust from the engine passes through an exhaust ATEG, the charge carriers of the semiconductors within the generator diffuse from the hot-side heat exchanger to the cold-side exchanger. The build-up of charge carriers results in a net charge, producing an electrostatic potential while the heat transfer drives a current. With exhaust temperatures of 700°C (~1300°F) or more, the temperature difference between exhaust gas on the hot side and coolant on the cold side is several hundred degrees. This temperature difference is capable of generating 500-750 W of electricity.

The compression assembly system aims to decrease the thermal contact resistance between the thermoelectric module and the heat exchanger surfaces. In coolant-based ATEGs, the cold side heat exchanger uses engine coolant as the cooling fluid, while in exhaust-based ATEGs, the cold-side heat exchanger uses ambient air as the cooling fluid.

Efficiency

Currently, ATEGs are about 5% efficient. However, advancements in thin-film and quantum well technologies could increase efficiency up to 15% in the future.

The efficiency of an ATEG is governed by the thermoelectric conversion efficiency of the materials and the thermal efficiency of the two heat exchangers. The ATEG efficiency can be expressed as:

$$\zeta_{OV} = \zeta_{CONV} \times \zeta_{HX} \times \rho$$

Where:

- ζ_{OV} : The overall efficiency of the ATEG

- ζ_{CONV} : Conversion efficiency of thermoelectric materials

- ζ_{HX}: Efficiency of the heat exchangers

- ρ : The ratio between the heat passed through thermoelectric materials to that passed from the hot side to the cold side

Benefits

The primary goal of ATEGs is to reduce fuel consumption and therefore reduce operating costs of a vehicle or help the vehicle comply with fuel efficiency standards.

Forty percent of an IC engine's energy is lost through exhaust gas heat. By converting the lost heat into electricity, ATEGs decrease fuel consumption by reducing the electric generator load on the engine. ATEGs allow the automobile to generate electricity from the engine's thermal energy rather than using mechanical energy to power an electric generator. Since the electricity is generated from waste heat that would otherwise be released into the environment, the engine burns less fuel to power the vehicle's electrical components, such as the headlights. Therefore, the automobile releases fewer emissions.

Decreased fuel consumption also results in increased fuel economy. Replacing the conventional electric generator with ATEGs could ultimately increase the fuel economy by up to 4%.

The ATEG's ability to generate electricity without moving parts is an advantage over mechanical electric generators alternatives.

Challenges

The greatest challenge to the scaling of ATEGs from prototyping to production has been the cost of the underlying thermoelectric materials. Since the early-2000's, many research agencies and institutions poured large sums of money into advancing the efficiency of thermoelectric materials. While efficiency improvements were made in materials such as the half heuslers and skutterudites, like their predecessors bismuth telluride and lead telluride, the cost of these materials has proven prohibitive for large-scale manufacturing. Recent advances by some researchers and companies in low-cost thermoelectric materials have resulted in significant commercial promise for ATEGs, most notably the invention of tetrahedrite by Michigan State University and its commercialization by US-based Alphabet Energy with General Motors.

Like any new component on an automobile, the use of an ATEG presents new engineering problems to consider, as well. However, given an ATEG's relatively low impact on the use of an automobile, its challenges are not as considerable as other new automotive technologies. For instance, since exhaust has to flow through the ATEG's heat exchanger, kinetic energy from the gas is lost, causing increased pumping losses. This is referred to as back pressure, which reduces the engine's performance. This can be accounted for by down-sizing the muffler, resulting in net-zero or even negative total back-pressure on the engine, as Faurecia and other companies have shown.

To make the ATEG's efficiency more consistent, coolant is usually used on the cold-side heat exchanger rather than ambient air so that the temperature difference will be the same on both hot and cold days. This may increases the radiator's size since piping must be extended to the exhaust manifold, and it may add to the radiator's load because there is more heat being transferred to the coolant. Proper thermal design does not require an up-sized cooling system.

ATEGs are made primarily of metal and, therefore, contribute more weight to the vehicle. The added weight causes the engine to work harder, resulting in lower gas mileage. Most automotive efficiency improvement studies of ATEGs, however, have resulted in a net positive efficiency gain even when considering the weight of the device.

History

Although the Seebeck effect was discovered in 1821, the use of thermoelectric power generators was restricted mainly to military and space applications until the second half of the twentieth century. This restriction was caused by the low conversion efficiency of thermoelectric materials at that time.

In 1963, the first ATEG was built and reported by Neild et al. In 1988, Birkholz et al. published the results of their work in collaboration with Porsche. These results described an exhaust-based ATEG which integrated iron-based thermoelectric materials between a carbon steel hot-side heat exchanger and an aluminium cold-side heat exchanger. This ATEG could produce tens of watts out of a Porsche 944 exhaust system.

In the early 1990s, Hi-Z Inc designed an ATEG which could produce 1 kW from a diesel truck exhaust system. The company in the following years introduced other designs for diesel trucks as well as military vehicles

In the late 1990s, Nissan Motors published the results of testing its ATEG which utilized SiGe thermoelectric materials. Nissan ATEG produced 35.6 W in testing conditions similar to the running conditions of a 3.0 L gasoline engine in hill-climb mode at 60.0 km/h.

Since the early-2000's, nearly every major automaker and exhaust supplier has experimented or studied thermoelectric generators, and companies including General Motors, BMW, Daimler, Ford, Renault, Honda, Toyota, Hyundai, Valeo, Boysen, Faurecia, Tenneco, Denso, Gentherm, Alphabet Energy, and numerous others have built and tested prototypes.

In January 2012, Car and Driver magazine named an ATEG created by a team led by Amerigon (now Gentherm Incorporated) one of the 10 "most promising" technologies.

Vibration-powered Generator

A vibration powered generator is a type of electric generator that converts the kinetic energy from vibration into electrical energy. The vibration may be from sound pressure waves or other ambient sources.

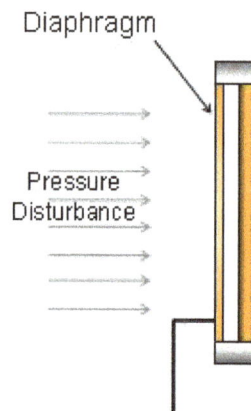

A transducer made of a piezoelectric diaphragm
being disturbed by a sound pressure wave.

Vibration powered generators usually consist of a resonator which is used to amplify the vibration source, and a transducer mechanism which converts the energy from the vibrations into electrical energy. The transducer usually consists of a magnet and coil or a piezoelectric crystal.

Electromagnetic Generators

Electromagnetic based generators use Faraday's law of induction to convert the kinetic energy of the vibrations into electrical energy. They consist of magnets attached to a flexible membrane or cantilever beam and a coil. The vibrations cause the distance between the magnet and coil to change, causing a change in magnetic flux and resulting in an electromagnetic force being produced. Generally the coil is made using a diamagnetic material as these materials have weaker interactions with the magnet that would dampen the vibration. The main advantage of this type of generator is that it is able to produce more power than the piezoelectric generators.

Development and Applications

A miniature electromagnetic vibration energy generator was developed by a team from the University of Southampton in 2007. This particular device consists of a cantilever beam with a magnet attached to the end. The beam moves up and down as the device is subjected to vibrations from surrounding sources. This device allows sensors in hard to access locations to be powered without electrical wires or batteries that need to be replaced. Sensors in inaccessible places can now generate their own power and transmit data to outside receivers. The generator was developed to be used in air compressors, and is able to power things in high vibration environments like sensors on machinery in manufacturing plants, or sensors that monitor the health of bridges. One of the major limitations of the magnetic vibration energy harvester developed at University of Southampton is the size of the generator. At approximately one cubic centimeter, this device would be much too large to be used in modern electronic devices. Future improvements on the size of the device could make it an ideal power source for medically implanted devices such as pacemakers. According to the team that created the device, the vibrations from the heart muscles would be enough to allow the generator to power a pacemaker. This would eliminate the need to replace the batteries surgically.

In 2012 a group at Northwestern University developed a vibration-powered generator out of polymer in the form of a spring. This device was able to harvest the energy from vibrations at the same frequencies as the University of Southampton groups cantilever based device, but at approximately one third the size of the other device.

Piezoelectric Generators

Piezoelectric based generators use thin membranes or cantilever beams made of piezoelectric crystals as a transducer mechanism. When the crystal is put under strain by the kinetic energy of the vibration a small amount of current is produced thanks to the piezoelectric effect. These mechanisms are usually very simple with few moving parts, and they tend to have a very long service life. This makes them the most popular method of harvesting the energy from vibrations. These mechanisms can be manufactured using the MEMS fabrication process, which allows them to be created on a very small scale. The ability to make piezoelectric generators on such a small scale is the main advantage of this method over the electromagnetic generators, especially when the generator is being developed to power microelectronic devices.

Development and Applications

One piezoelectric generator being developed uses water droplets at the end of a piezoelectric cantilever beam. The water droplets hang from the end of the beam and are subjected to excitation by the kinetic energy of the vibrations. This results in the water droplet oscillating, which in turn causes the beam they are hanging from to deflect up and down. This deflection is the strain which is converted to energy through the piezoelectric effect. A major advantage to this method is that it can be tailored towards a wide range of excitation frequencies. The natural frequency of the water droplet is a function of its size; therefore changing the size of the water droplet allows for the matching of the natural frequency of the droplet and the frequency of the pressure wave being converted into electrical energy. Matching these frequencies produces the largest amplitude oscillation of the water droplet, resulting in a large force and larger strain on the piezoelectric beam.

Another application seeks to use the vibrations created during flight in aircraft to power the electronics on the plane that currently rely on batteries. Such a system would allow for a reliable energy source, and reduce maintenance as batteries would no longer need to be replaced and piezoelectric systems have a long service life. This system is used with a resonator, which allows the airflow to form a high amplitude steady tone. The same principle is used in many wind instruments, converting the airflow provided by the musician into a loud steady tone. This tone is used as the vibration that is converted from kinetic to electric energy by the piezoelectric generator. This application is still in the early stages of development; the concept has been proven on a scale model but the system still needs to be optimized before it is tested on a full scale.

Rectenna

A rectenna is a rectifying antenna—a special type of antenna that is used for converting electromagnetic energy into direct current (DC) electricity. They are used in wireless power transmission systems that transmit power by radio waves. A simple rectenna element consists of a dipole antenna with an RF diode connected across the dipole elements. The diode rectifies the AC current induced in the antenna by the microwaves, to produce DC power, which powers a load connected across the diode. Schottky diodes are usually used because they have the lowest voltage drop and highest speed and therefore have the lowest power losses due to conduction and switching. Large rectennas consist of an array of many such dipole elements.

The invention of the rectenna in the 1960s made long distance wireless power transmission feasible. The rectenna was invented in 1964 and patented in 1969 by US electrical engineer William C. Brown, who demonstrated it with a model helicopter powered by microwaves transmitted from the ground, received by an attached rectenna. Since the 1970s, one of the major motivations for rectenna research has been to develop a receiving antenna for proposed solar power satellites, which would harvest energy from sunlight in space with solar cells and beam it down to Earth as microwaves to huge rectenna arrays. A proposed military application is to power drone reconnaissance aircraft with microwaves beamed from the ground, allowing them to stay aloft for long periods. In recent years, interest has turned to using rectennas as power sources for small wireless microelectronic devices. The largest current use of rectennas is in RFID tags, proximity cards and contactless smart cards, which contain an integrated circuit (IC) which is powered by a small rectenna element. When the device is brought near an electronic reader unit, radio waves from the reader are received by the rectenna, powering up the IC, which transmits its data back to the reader.

Radio Frequency Rectennas

The simplest crystal radio receiver, employing an antenna and a demodulating diode (rectifier), is actually a rectenna - although it discards the DC component before sending the signal to the earphones. People living near strong radio transmitters would occasionally discover that with a long receiving antenna, they could get enough electric power to light a light bulb.

However this example uses only one antenna having a limited capture area. A Rectenna uses multiple antennas spread over a wide area to capture more energy.

Researchers are experimenting with the use of rectennas to power sensors in remote areas.

RF rectennas are used for several forms of wireless energy transfer. In the microwave range, experimental devices already reached a power conversion efficiency of 85-90%.

Optical Rectennas

Similar devices, scaled down to the proportions used in nanotechnology, can be used to convert light directly into electricity. This type of device is called an *optical rectenna* or *nantenna*. Theoretically, high efficiencies can be maintained as the device shrinks, but efficiency has so far been limited. The University of Missouri previously reported on work to develop low-cost, high-efficiency nantennas (optical-frequency rectennas). Other prototype devices were investigated in a collaboration between the University of Connecticut and Penn State Altoona using a grant from the National Science Foundation. With the use of atomic layer deposition it has been suggested that conversion efficiencies of solar energy to electricity higher than 70% could eventually be achieved.

Challenges to successful nantenna technology include fabricating an antenna small enough to couple optical wavelengths, and creating an ultrafast diode capable of rectifying the high frequency oscillations.

Carbon Nanotube Optical Rectenna

In 2015, researchers at Georgia Institute of Technology fabricated an optical rectenna using arrays of 2 million multiwall carbon nanotubes (MWCNT) per cm^2 coupled to nanoscale rectifying diodes. The MWCNT, which act as optical antennae due to their favorably small dimensions, are coated in aluminum oxide and capped with a metallic top layer. This combination of MWCNT, oxide, metal is claimed to be the world's fastest metal-insulator-metal (MIM) tunneling diode, capable of rectifying optical frequencies. Individual junctions were reported to have a capacitance of only 1.7 attofarads, with switching time on the order of 1 femtosecond.

Windbelt

The Windbelt is a wind power harvesting device invented by Shawn Frayn in 2004 for converting wind power to electricity. It consists of a flexible polymer ribbon stretched between supports transverse to the wind direction, with magnets glued to it. When the wind blows across it, the ribbon vibrates due to aeroelastic flutter, similar to the action of an aeolian harp. The vibrating movement of the magnets induces current in nearby pickup coils by electromagnetic induction.

One prototype has powered two LEDs, a radio, and a clock (separately) using wind generated from a household fan. The cost of the materials was well under US$10. $2–$5 for 40 mW is a cost of $50–$125 per watt.

There are three sizes in development:

- The microBelt, a 12 cm version. This could be put into production in around six months. Its expected to produce 1 milliwatt average. To charge a pair of ideal rechargeable AA cells (2.5Ah 1.2v) this would take 6000 hours, or 250 days.

- The Windcell, a 1-metre version that could be used to power meshed WiFi repeaters, charge cellphones, or run LED lights. This could go into production within 18 to 24 months. It is hoped that a square metre panel at 6 m/s average windspeed can generate 10 W average.

- An experimental 10-metre model that has no production date.

The Windbelt's inventor, Shawn Frayn, was a winner of the 2007 Breakthrough Award from the publishers of the magazine, *Popular Mechanics*. He is trying to make the Windbelt cheaper.

The inventor's claims that the device is 10 - 30 times more efficient than small wind turbines have been refuted by tests. The microWindbelt could generate 0.2 mW at a wind speed of 3.5 m/s and 5 mW at 7.5 m/s, which represent efficiencies (ηC_p) of 0.21% and 0.53% respectively. Wind turbines typically have efficiencies of 1% to 10%. Since the Windbelt a number of other "flutter" wind harvester devices have been designed, but like the Windbelt almost all have efficiencies below turbine machines.

Solar-assisted Heat Pump

A solar-assisted heat pump (SAHP) is a machine that represents the integration of a heat pump and thermal solar panels in a single integrated system. Typically these two technologies are used separately (or only placing them in parallel) to produce hot water. In this system the solar thermal panel performs the function of the low temperature heat source and the heat produced is used to feed the heat pump's evaporator. The goal of this system is to get high COP and then produce energy in a more efficient and less expensive way.

It is possible to use any type of solar thermal panel (sheet and tubes, roll-bond, heat pipe, thermal plates) or hybrid (mono/polycrystalline, thin film) in combination with the heat pump. The use of a hybrid panel is preferable because it allows to cover a part of the electricity demand of the heat pump and reduce the power consumption and consequently the variable costs of the system.

Hybrid photovoltaic-thermal solar panels of a SAHP in an experimental installation at Department of Energy at Polytechnic of Milan.

Optimization

The operating conditions' optimization of this system is the main problem, because there are two opposing trends of the performance of the two sub-systems: by way of example, a decreasing of the evaporation temperature of the working fluid generates an increasing of the thermal efficiency of the solar panel but a decreasing in the performance of the heat pump, with a decreasing in the COP. The target for the optimization is normally the minimization of the electrical consumption of the heat pump, or primary energy required by an auxiliary boiler which supplies the load not covered by renewable source.

Configurations

There are two possible configurations of this system, which are distinguished by the presence or not of an intermediate fluid that transports the heat from the panel to the heat pump. Machines called indirect- expansion mainly use water as a heat transfer fluid, mixed with an antifreeze fluid (usually glycol) to avoid ice formation phenomena during winter period. The machines called direct-expansion put in the refrigerant fluid directly inside the hydraulic circuit of the thermal panel, where the phase transition takes place. This second configuration, even though it is more complex from the technical point of view, allows to obtain several advantages:

- a better transfer of the heat produced by the thermal panel to the working fluid which involves a greater thermal efficiency of the evaporator, linked to the absence of an intermediate fluid;

- presence of an evaporating fluid allows to obtain a uniform temperature distribution in the thermal panel with a consequent increase in the thermal efficiency (in normal operating conditions of the solar panel, the local thermal efficiency decreases from inlet to outlet of the fluid because the fluid temperature increases);

- using hybrid solar panel, in addition to the advantage described in the previous point, the electrical efficiency of the panel increases (for similar considerations).

Comparison

Generally speaking the use of this integrated system is an efficient way to employ the heat produced by the thermal panels in winter period, which wouldn't be normally exploited because of its too low temperature.

Separated Production Systems

In comparison with only heat pump utilization, it is possible to reduce the amount of electrical energy consumed by the machine during the weather evolution from winter season to the spring, and then finally only use thermal solar panels to produce all the heat demand required (only in case of indirect-expansion machine), thus saving on variable costs.

In comparison with a system with only thermal panels, it is possible to supply to a greater part of the winter heating required using a non-fossil energy source.

Traditional Heat Pumps

Compared to geothermal heat pumps, the main advantage is that the installation of a piping field in the soil is not required, which results in a lower cost of investment (in the investment cost of a geothermal heat pump system about 50% is given by drilling costs) and in more flexibility of machine installation, even in areas in which there is poor space availability. Furthermore, there are no risks related to possible thermal soil impoverishment. A disadvantage, similarly to as occurs for air heat pumps, solar-assisted heat pump performances are affected by atmospheric conditions.

Compared to air heat pumps, as already mentioned, there are variations less significant of the performances (linked to the variation of solar radiation and not to air temperature oscillation). This produces a greater SCOP (Seasonal COP). In addition to this, the evaporation temperature of the working fluid is higher than the case for air, so in general the coefficient of performance is significantly higher.

Low Temperature Conditions

In general, a heat pump can evaporate at temperatures below the ambient temperature. In a solar-assisted heat pump this generates a temperature distribution of the thermal panels below that temperature. In this condition thermal losses of the panels towards the environment become additional available energy to the heat pump. In this case it is possible that the thermal efficiency of solar panels is more than 100%.

Another free-contribution in these conditions of low temperature is related to the possibility of condensation of water vapor on the surface of the panels, which provides additional heat to the heat transfer fluid (normally it is a small part of the total heat collected by solar panels), that is equal to the latent heat of condensation.

Heat Pump with Double Cold Sources

The simple configuration of solar-assisted heat pump as only solar panels as heat source for the evaporator. It can also exist a configuration with an additional heat source. The goal is to have further advantages in energy saving but, in the other hand, the management and optimization of the system become more complex.

The geothermal-solar configuration allows to reduce the size of the piping field (and reduce the investment) and to have a regeneration of the ground during summer through the heat collected from the thermal panels.

The air-solar structure allows to have an acceptable heat input also during cloudy days, maintaining the compactness of the system and the easiness to install it.

References

- Moubarak, P.; et al. (2012). "A Self-Calibrating Mathematical Model for the Direct Piezoelectric Effect of a New MEMS Tilt Sensor". IEEE Sensors Journal. 12 (5): 1033–1042. doi:10.1109/jsen.2011.2173188

- Hau, Erich. "Wind Turbines: Fundamentals, Technologies, Application, Economics" p142. Springer Science & Business Media, 26. feb. 2013. ISBN 3642271510

- Text und Photos: ENERCON Germany www.enercon.de. "Anatomy of an Enercon direct drive wind turbine". Wwindea.org. Retrieved 2013-11-06

- Christou, P (2007). "Advanced materials for turbine blade manufacture". Reinforced Plastics. 51 (4): 22. doi:10.1016/S0034-3617(07)70148-0

- Kim, D.S. (January 2008). "Solar refrigeration options – a state-of-the-art review". International Journal of Refrigeration. 31 (1): 3–15. doi:10.1016/j.ijrefrig.2007.07.011. Retrieved 8 November 2016

- Zbigniew Lubosny (2003). Wind Turbine Operation in Electric Power Systems: Advanced Modeling (Power Systems). Berlin: Springer. ISBN 3-540-40340-X

- "Materials and design methods look for the 100-m blade". Windpower Engineering. 10 May 2011. Retrieved 22 August 2011

- Ali M. Eltamaly, A. I. Alolah, and Hassan M. Farh. "Maximum Power Extraction from Utility-Interfaced Wind Turbines". 2013. DOI: 10.5772/54675

- Yang, Jihui. "Automotive Applications of Thermoelectric Materials". Journal of Electronic Materials, 2009, VOL 38; page 1245

- Holler, F. James; Skoog, Douglas A.; Crouch, Stanley R. (2007). "Chapter 1". Principles of Instrumental Analysis (6th ed.). Cengage Learning. p. 9. ISBN 978-0-495-01201-6

- McGar, Justin. "Wind Power Revolution: The World's First Timber Turbine" Design Build Source, 13 November 2012. Retrieved: 13 November 2012

- Arjun, A.; Sampath, A.; Thiyagarajan, S.; Arvind, V. (December 2011). "A Novel Approach to Recycle Energy Using Piezoelectric Crystals". International Journal of Environmental Science and Development. 2: 488–492. doi:10.7763/IJESD.2011.V2.175

Photovoltaics: An Integrated Study

Photovoltaics is the term used for the transformation of light into electricity. The process is accomplished by using semiconducting materials. Concentrator photovoltaics, photovoltaic thermal hybrid solar collector, solar cell, solar cell efficiency and solar panels on spacecraft are some important topics related to the subject of photovoltaics. This chapter will provide an integrated understanding of photovoltaics.

Photovoltaics

The Solar Settlement, a sustainable housing community project in Freiburg, Germany.

Photovoltaics (PV) is a term which covers the conversion of light into electricity using semiconducting materials that exhibit the photovoltaic effect, a phenomenon studied in physics, photochemistry, and electrochemistry.

A typical photovoltaic system employs solar panels, each comprising a number of solar cells, which generate electrical power. PV installations may be ground-mounted, rooftop mounted or wall mounted. The mount may be fixed, or use a solar tracker to follow the sun across the sky.

Solar PV has specific advantages as an energy source: its operation generates no pollution and no greenhouse gas emissions once installed, it shows simple scalability in respect of power needs and silicon has large availability in the Earth's crust.

PV systems have the major disadvantage that the power output is dependent on direct sunlight, so about 10-25% is lost if a tracking system is not used, since the cell will not be directly facing the sun at all times. Dust, clouds, and other things in the atmosphere also diminish the power output. Another main issue is the concentration of the production in the hours corresponding to main insolation, which don't usually match the peaks in demand in human activity cycles. Unless current societal patterns of consumption and electrical networks mutually adjust to this scenario, electricity still needs to be made up by other power sources, usually hydrocarbon.

Photovoltaic SUDI shade is an autonomous and mobile station in France that provides energy for electric vehicles using solar energy.

Solar panels on the International Space Station

Photovoltaic systems have long been used in specialized applications, and standalone and grid-connected PV systems have been in use since the 1990s. They were first mass-produced in 2000, when German environmentalists and the Eurosolar organization got government funding for a ten thousand roof program.

Advances in technology and increased manufacturing scale have in any case reduced the cost, increased the reliability, and increased the efficiency of photovoltaic installations. Net metering and financial incentives, such as preferential feed-in tariffs for solar-generated electricity, have supported solar PV installations in many countries. More than 100 countries now use solar PV.

After hydro and wind powers, PV is the third renewable energy source in terms of globally capacity. In 2014, worldwide installed PV capacity increased to 177 gigawatts (GW), which is two percent of global electricity demand. China, followed by Japan and the United States, is the fastest growing market, while Germany remains the world's largest producer, with solar PV providing seven percent of annual domestic electricity consumption. With current technology (as of 2013), photovoltaics recoups the energy needed to manufacture them in 1.5 years in Southern Europe and 2.5 years in Northern Europe.

Etymology

The term "photovoltaic" comes from the Greek (*phōs*) meaning "light", and from "volt", the unit of electro-motive force, the volt, which in turn comes from the last name of the Italian physicist Alessandro Volta, inventor of the battery (electrochemical cell). The term "photo-voltaic" has been in use in English since 1849.

Solar Cells

Solar cells generate electricity directly from sunlight.

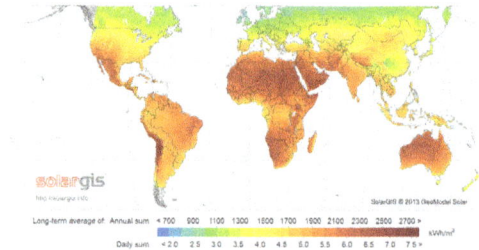

Average insolation. Note that this is for a horizontal surface. Solar panels are normally propped up at an angle and receive more energy per unit area.

Photovoltaics are best known as a method for generating electric power by using solar cells to convert energy from the sun into a flow of electrons by the photovoltaic effect.

Solar cells produce direct current electricity from sunlight which can be used to power equipment or to recharge a battery. The first practical application of photovoltaics was to power orbiting satellites and other spacecraft, but today the majority of photovoltaic modules are used for grid connected power generation. In this case an inverter is required to convert the DC to AC. There is a smaller market for off-grid power for remote dwellings, boats, recreational vehicles, electric cars, roadside emergency telephones, remote sensing, and cathodic protection of pipelines.

Photovoltaic power generation employs solar panels composed of a number of solar cells containing a photovoltaic material. Copper solar cables connect modules (module cable), arrays (array cable), and sub-fields. Because of the growing demand for renewable energy sources, the manufacturing of solar cells and photovoltaic arrays has advanced considerably in recent years.

Solar photovoltaic power generation has long been seen as a clean energy technology which draws upon the planet's most plentiful and widely distributed renewable energy source – the sun. Cells require protection from the environment and are usually packaged tightly in solar panels.

Photovoltaic power capacity is measured as maximum power output under standardized test conditions (STC) in "W_p" (watts peak). The actual power output at a particular point in time may be less than or greater than this standardized, or "rated," value, depending on geographical location, time of day, weather conditions, and other factors. Solar photovoltaic array capacity factors are typically under 25%, which is lower than many other industrial sources of electricity.

Current Developments

For best performance, terrestrial PV systems aim to maximize the time they face the sun. Solar trackers achieve this by moving PV panels to follow the sun. The increase can be by as much as 20% in winter and by as much as 50% in summer. Static mounted systems can be optimized by analysis of the sun path. Panels are often set to latitude tilt, an angle equal to the latitude, but performance can be improved by adjusting the angle for summer or winter. Generally, as with other semiconductor devices, temperatures above room temperature reduce the performance of photovoltaics.

A number of solar panels may also be mounted vertically above each other in a tower, if the zenith distance of the Sun is greater than zero, and the tower can be turned horizontally as a whole and each panels additionally around a horizontal axis. In such a tower the panels can follow the Sun exactly. Such a device may be described as a ladder mounted on a turnable disk. Each step of that ladder is the middle axis of a rectangular solar panel. In case the zenith distance of the Sun reaches zero, the "ladder" may be rotated to the north or the south to avoid a solar panel producing a shadow on a lower solar panel. Instead of an exactly vertical tower one can choose a tower with an axis directed to the polar star, meaning that it is parallel to the rotation axis of the Earth. In this case the angle between the axis and the Sun is always larger than 66 degrees. During a day it is only necessary to turn the panels around this axis to follow the Sun. Installations may be ground-mounted (and sometimes integrated with farming and grazing) or built into the roof or walls of a building (building-integrated photovoltaics).

Another recent development involves the makeup of solar cells. Perovskite is a very inexpensive material which is being used to replace the expensive crystalline silicon which is still part of a standard PV cell build to this day. Michael Graetzel, Director of the Laboratory of Photonics and Interfaces at EPFL says, "Today, efficiency has peaked at 18 percent, but it's expected to get even higher in the future." This is a significant claim, as 20% efficiency is typical among solar panels which use more expensive materials.

Efficiency

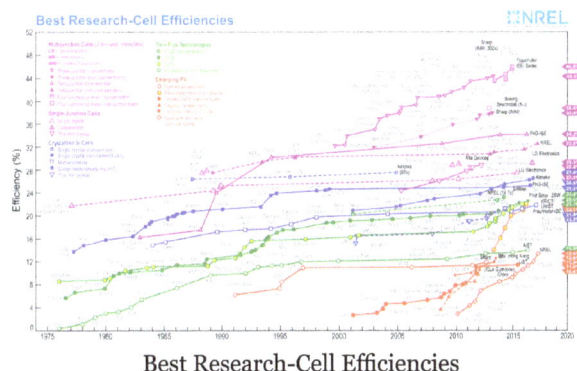

Best Research-Cell Efficiencies

Electrical efficiency (also called conversion efficiency) is a contributing factor in the selection of a photovoltaic system. However, the most efficient solar panels are typically the most expensive, and may not be commercially available. Therefore, selection is also driven by cost efficiency and other factors.

The electrical efficiency of a PV cell is a physical property which represents how much electrical power a cell can produce for a given insolation. The basic expression for maximum efficiency of a photovoltaic cell is given by the ratio of output power to the incident solar power (radiation flux times area)

$$\eta = \frac{P_{max}}{E \cdot A_{cell}}.$$

The efficiency is measured under ideal laboratory conditions and represents the maximum achievable efficiency of the PV material. Actual efficiency is influenced by the output Voltage, current, junction temperature, light intensity and spectrum.

The most efficient type of solar cell to date is a multi-junction concentrator solar cell with an efficiency of 46.0% produced by Fraunhofer ISE in December 2014. The highest efficiencies achieved without concentration include a material by Sharp Corporation at 35.8% using a proprietary triple-junction manufacturing technology in 2009, and Boeing Spectrolab (40.7% also using a triple-layer design). The US company SunPower produces cells that have an efficiency of 21.5%, well above the market average of 12–18%.

There is an ongoing effort to increase the conversion efficiency of PV cells and modules, primarily for competitive advantage. In order to increase the efficiency of solar cells, it is important to choose a semiconductor material with an appropriate band gap that matches the solar spectrum. This will enhance the electrical and optical properties. Improving the method of charge collection is also useful for increasing the efficiency. There are several groups of materials that are being developed. Ultrahigh-efficiency devices ($\eta > 30\%$) are made by using GaAs and GaInP2 semiconductors with multijunction tandem cells. High-quality, single-crystal silicon materials are used to achieve high-efficiency, low cost cells ($\eta > 20\%$).

Recent developments in Organic photovoltaic cells (OPVs) have made significant advancements in power conversion efficiency from 3% to over 15% since their introduction in the 1980s. To date, the highest reported power conversion efficiency ranges from 6.7% to 8.94% for small molecule, 8.4%–10.6% for polymer OPVs, and 7% to 21% for perovskite OPVs. OPVs are expected to play a major role in the PV market. Recent improvements have increased the efficiency and lowered cost, while remaining environmentally-benign and renewable.

Several companies have begun embedding power optimizers into PV modules called smart modules. These modules perform maximum power point tracking (MPPT) for each module individually, measure performance data for monitoring, and provide additional safety features. Such modules can also compensate for shading effects, wherein a shadow falling across a section of a module causes the electrical output of one or more strings of cells in the module to decrease.

One of the major causes for the decreased performance of cells is overheating. The efficiency of a solar cell declines by about 0.5% for every 1 degree Celsius increase in temperature. This means that a

100 degree increase in surface temperature could decrease the efficiency of a solar cell by about half. Self-cooling solar cells are one solution to this problem. Rather than using energy to cool the surface, pyramid and cone shapes can be formed from silica, and attached to the surface of a solar panel. Doing so allows visible light to reach the solar cells, but reflects infrared rays (which carry heat).

Growth

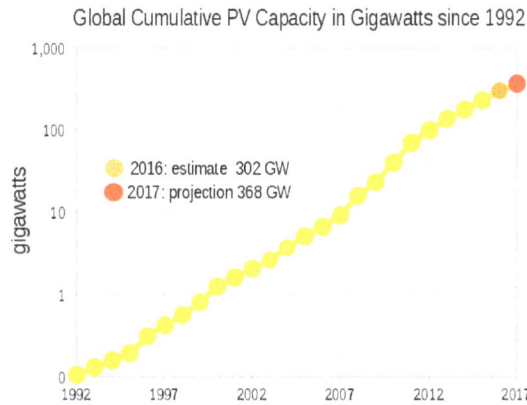

Worldwide growth of photovoltaics on a semi-log plot since 1992

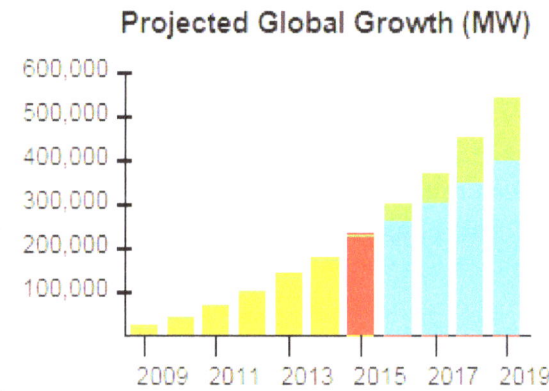

Projected global cumulative capacity (SPE)

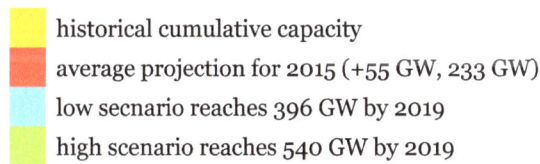

- historical cumulative capacity
- average projection for 2015 (+55 GW, 233 GW)
- low secnario reaches 396 GW by 2019
- high scenario reaches 540 GW by 2019

Solar photovoltaics is growing rapidly and worldwide installed capacity reached about 227 gigawatts (GW) by the end of 2015. Since 2000, installed capacity has seen an growth factor of about 57. The total power output of the world's PV capacity in a calendar year in 2014 is now beyond 200 TWh of electricity. This represents 1% of worldwide electricity demand. More than 100 countries use solar PV. China, followed by Japan and the United States is now the fastest growing market, while Germany remains the world's largest producer, contributing more than 7% to its national electricity demands. Photovoltaics is now, after hydro and wind power, the third most important renewable energy source in terms of globally installed capacity.

Several market research and financial companies foresee record-breaking global installation of more than 50 GW in 2015. China is predicted to take the lead from Germany and to become the world's largest producer of PV power by installing another targeted 17.8 GW in 2015. India is expected to install 1.8 GW, doubling its annual installations. By 2018, worldwide photovoltaic capacity is projected to doubled or even triple to 430 GW. Solar Power Europe (formerly known as EPIA) also estimates that photovoltaics will meet 10% to 15% of Europe's energy demand in 2030.

In 2017 a study in Science estimated that by 2030 global PV installed capacities will be between 3,000 and 10,000 GW. The EPIA/Greenpeace Solar Generation Paradigm Shift Scenario (formerly called Advanced Scenario) from 2010 shows that by the year 2030, 1,845 GW of PV systems could be generating approximately 2,646 TWh/year of electricity around the world. Combined with energy use efficiency improvements, this would represent the electricity needs of more than 9% of the world's population. By 2050, over 20% of all electricity could be provided by photovoltaics.

Michael Liebreich, from Bloomberg New Energy Finance, anticipates a tipping point for solar energy. The costs of power from wind and solar are already below those of conventional electricity generation in some parts of the world, as they have fallen sharply and will continue to do so. He also asserts, that the electrical grid has been greatly expanded worldwide, and is ready to receive and distribute electricity from renewable sources. In addition, worldwide electricity prices came under strong pressure from renewable energy sources, that are, in part, enthusiastically embraced by consumers.

Deutsche Bank sees a "second gold rush" for the photovoltaic industry to come. Grid parity has already been reached in at least 19 markets by January 2014. Photovoltaics will prevail beyond feed-in tariffs, becoming more competitive as deployment increases and prices continue to fall.

In June 2014 Barclays downgraded bonds of U.S. utility companies. Barclays expects more competition by a growing self-consumption due to a combination of decentralized PV-systems and residential electricity storage. This could fundamentally change the utility's business model and transform the system over the next ten years, as prices for these systems are predicted to fall.

Top 10 PV countries in 2015 (MW)			
Installed and Total Solar Power Capacity in 2015 (MW)			
#	Nation	Total Capacity	Added Capacity
1	China	43,530	15,150
2	Germany	39,700	1,450
3	Japan	34,410	11,000
4	United States	25,620	7,300
5	Italy	18,920	300
6	United Kingdom	8,780	3,510
7	France	6,580	879
8	Spain	5,400	56
9	Australia	5,070	935
10	India	5,050	2,000

Environmental Impacts of Photovoltaic Technologies

Types of impacts

While solar photovoltaic (PV) cells are promising for clean energy production, their deployment is hindered by production costs, material availability, and toxicity. Data required to investigate their impact are sometimes affected by a rather big amount of uncertainty. The values of human labor and water consumption, for example, are not precisely assessed due to the lack of systematic and accurate analyses in the scientific literature.

Life cycle assessment (LCA) is one method of determining environmental impacts from PV. Many studies have been done on the various types of PV including first generation, second generation, and third generation. Usually these PV LCA studies select a cradle to gate system boundary because often at the time the studies are conducted, it is a new technology not commercially available yet and their required balance of system components and disposal methods are unknown.

A traditional LCA can look at many different impact categories ranging from global warming potential, eco-toxicity, human toxicity, water depletion, and many others.

Most LCAs of PV have focused on two categories: carbon dioxide equivalents per kWh and energy pay-back time (EPBT). The EPBT is defined as " the time needed to compensate for the total renewable- and non-renewable- primary energy required during the life cycle of a PV system". A 2015 review of EPBT from first and second generation PV suggested that there was greater variation in embedded energy than in efficiency of the cells implying that it was mainly the embedded energy that needs to reduce to have a greater reduction in EPBT. One difficulty in determining impacts due to PV is to determine if the wastes are released to the air, water, or soil during the manufacturing phase. Research is underway to try to understand emissions and releases during the lifetime of PV systems.

Impacts from first-generation PV

Crystalline silicon modules are the most extensively studied PV type in terms of LCA since they are the most commonly used. Mono-crystalline silicon photovoltaic systems (mono-si) have an average efficiency of 14.0%. The cells tend to follow a structure of front electrode, anti-reflection film, n-layer, p-layer, and back electrode, with the sun hitting the front electrode. EPBT ranges from 1.7 to 2.7 years. The cradle to gate of CO_2-eq/kWh ranges from 37.3 to 72.2 grams.

Techniques to produce multi-crystalline silicon (multi-si) photovoltaic cells are simpler and cheaper than mono-si, however tend to make less efficient cells, an average of 13.2%. EPBT ranges from 1.5 to 2.6 years. The cradle to gate of CO_2-eq/kWh ranges from 28.5 to 69 grams. Some studies have looked beyond EPBT and GWP to other environmental impacts. In one such study, conventional energy mix in Greece was compared to multi-si PV and found a 95% overall reduction in impacts including carcinogens, eco-toxicity, acidification, eutrophication, and eleven others.

Impacts from second generation

Cadmium telluride (CdTe) is one of the fastest-growing thin film based solar cells which are collectively known as second generation devices. This new thin film device also shares similar performance restrictions (Shockley-Queisser efficiency limit) as conventional Si devices but promises to lower the cost of each device by both reducing material and energy consumption during manufacturing. Today the global market share of CdTe is 5.4%, up from 4.7% in 2008. This technology's highest power conversion efficiency is 21%. The cell structure includes glass substrate (around 2 mm), transparent conductor layer, CdS buffer layer (50–150 nm), CdTe absorber and a metal contact layer.

CdTe PV systems require less energy input in their production than other commercial PV systems per unit electricity production. The average CO_2-eq/kWh is around 18 grams (cradle to gate). CdTe has the fastest EPBT of all commercial PV technologies, which varies between 0.3 and 1.2 years.

Copper Indium Gallium Diselenide (CIGS) is a thin film solar cell based on the copper indium diselenide (CIS) family of chalcopyrite semiconductors. CIS and CIGS are often used interchangeably within the CIS/CIGS community. The cell structure includes soda lime glass as the substrate, Mo layer as the back contact, CIS/CIGS as the absorber layer, cadmium sulfide (CdS) or Zn (S,OH) x as the buffer layer, and ZnO:Al as the front contact. CIGS is approximately 1/100th the thickness of conventional silicon solar cell technologies. Materials necessary for assembly are readily available, and are less costly per watt of solar cell. CIGS based solar devices resist performance degradation over time and are highly stable in the field.

Reported global warming potential impacts of CIGS range from 20.5 – 58.8 grams CO_2-eq/kWh of electricity generated for different solar irradiation (1,700 to 2,200 kWh/m²/y) and power conversion efficiency (7.8 – 9.12%). EPBT ranges from 0.2 to 1.4 years, while harmonized value of EPBT was found 1.393 years. Toxicity is an issue within the buffer layer of CIGS modules because it contains cadmium and gallium. CIS modules do not contain any heavy metals.

Impacts from third generation

Third-generation PVs are designed to combine the advantages of both the first and second generation devices and they do not have Shockley-Queisser limit, a theoretical limit for first and second generation PV cells. The thickness of a third generation device is less than 1 μm.

One emerging alternative and promising technology is based on an organic-inorganic hybrid solar cell made of methylammonium lead halide perovskites. Perovskite PV cells have progressed rapidly over the past few years and have become one of the most attractive areas for PV research. The cell structure includes a metal back contact (which can be made of Al, Au or Ag), a hole transfer layer (spiro-MeOTAD, P3HT, PTAA, CuSCN, CuI, or NiO), and absorber layer ($CH_3NH_3PbIxBr_3$-x, $CH_3NH_3PbIxCl_3$-x or $CH_3NH_3PbI_3$), an electron transport layer (TiO, ZnO, Al_2O_3 or SnO_2) and a top contact layer (fluorine doped tin oxide or tin doped indium oxide).

There are a limited number of published studies to address the environmental impacts of perovskite solar cells. The major environmental concern is the lead used in the absorber layer. Due to the instability of perovskite cells lead may eventually be exposed to fresh water during the use phase. These LCA studies looked at human and ecotoxicity of perovskite solar cells and found they were surprisingly low and may not be an environmental issue. Global warming potential of perovskite PVs were found to be in the range of 24–1500 grams CO_2-eq/kWh electricity production. Similarly, reported EPBT of the published paper range from 0.2 to 15 years. The large range of reported values highlight the uncertainties associated with these studies. Celik et al. (2016) critically discussed the assumptions made in perovskite PV LCA studies.

Two new promising thin film technologies are copper zinc tin sulfide (Cu_2ZnSnS_4 or CZTS), zinc phosphide (Zn_3P_2) and single-walled carbon nano-tubes (SWCNT). These thin films are currently only produced in the lab but may be commercialized in the future. The manufacturing of CZTS and (Zn_3P_2) processes are expected to be similar to those of current thin film technologies of CIGS and CdTe, respectively. While the absorber layer of SWCNT PV is expected to be synthesized with CoMoCAT method. by Contrary to established thin films such as CIGS and CdTe, CZTS, Zn_3P_2, and SWCNT PVs are made from earth abundant, nontoxic materials and have the potential to produce more electricity annually than the current worldwide consumption. While CZTS and Zn_3P_2 offer good promise for these reasons, the specific environmental implications of their commercial production are not yet known. Global warming potential of CZTS and Zn_3P_2 were found 38 and 30 grams CO_2-eq/kWh while their corresponding EPBT were found 1.85 and 0.78 years, respectively. Overall, CdTe and Zn_3P_2 have similar environmental impacts but can slightly outperform CIGS and CZTS. Celik et al. performed the first LCA study on environmental impacts of SWCNT PVs, including a laboratory-made 1% efficient device and an aspirational 28% efficient four-cell tandem device and interpreted the results by using mono-Si as a reference point. the results show that compared to monocrystalline Si (mono-Si), the environmental impacts from 1% SWCNT was ~18 times higher due mainly to the short lifetime of three years. However, even with the same short lifetime, the 28% cell had lower environmental impacts than mono-Si.

Organic and polymer photovoltaic (OPV) are a relatively new area of research. The tradition OPV cell structure layers consist of a semi-transparent electrode, electron blocking layer, tunnel junction, holes blocking layer, electrode, with the sun hitting the transparent electrode. OPV replaces silver with carbon as an electrode material lowering manufacturing cost and making them more environmentally friendly. OPV are flexible, low weight, and work well with roll-to roll manufacturing for mass production. OPV uses "only abundant elements coupled to an extremely low embodied energy through very low processing temperatures using only ambient processing conditions on simple printing equipment enabling energy pay-back times". Current efficiencies range from 1–6.5%, however theoretical analyses show promise beyond 10% efficiency.

Many different configurations of OPV exist using different materials for each layer. OPV technology rivals existing PV technologies in terms of EPBT even if they currently present a shorter operational lifetime. A 2013 study analyzed 12 different configurations all with 2% efficiency, the EPBT ranged from 0.29–0.52 years for 1 m² of PV. The average CO_2-eq/kWh for OPV is 54.922 grams.

Economics

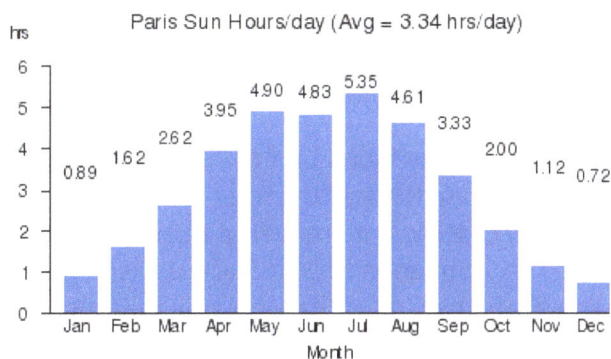

Paris Sun Hours/day (Avg = 3.34 hrs/day)

Source: Apricus

There have been major changes in the underlying costs, industry structure and market prices of solar photovoltaics technology, over the years, and gaining a coherent picture of the shifts occurring across the industry value chain globally is a challenge. This is due to: "the rapidity of cost and price changes, the complexity of the PV supply chain, which involves a large number of manufacturing processes, the balance of system (BOS) and installation costs associated with complete PV systems, the choice of different distribution channels, and differences between regional markets within which PV is being deployed". Further complexities result from the many different policy support initiatives that have been put in place to facilitate photovoltaics commercialisation in various countries.

The PV industry has seen dramatic drops in module prices since 2008. In late 2011, factory-gate prices for crystalline-silicon photovoltaic modules dropped below the $1.00/W mark. The $1.00/W installed cost, is often regarded in the PV industry as marking the achievement of grid parity for PV. Technological advancements, manufacturing process improvements, and industry re-structuring, mean that further price reductions are likely in coming years. As of 2017 power-purchase agreement prices for solar farms below $0.05/kWh are common in the United States and the lowest bids in several international countries were about $0.03/kWh.

Financial incentives for photovoltaics, such as feed-in tariffs, have often been offered to electricity consumers to install and operate solar-electric generating systems. Government has sometimes also offered incentives in order to encourage the PV industry to achieve the economies of scale needed to compete where the cost of PV-generated electricity is above the cost from the existing grid. Such policies are implemented to promote national or territorial energy independence, high tech job creation and reduction of carbon dioxide emissions which cause global warming. Due to economies of scale solar panels get less costly as people use and buy more—as manufacturers increase production to meet demand, the cost and price is expected to drop in the years to come.

Solar cell efficiencies vary from 6% for amorphous silicon-based solar cells to 44.0% with multiple-junction concentrated photovoltaics. Solar cell energy conversion efficiencies for commercially available photovoltaics are around 14–22%. Concentrated photovoltaics (CPV) may reduce cost by concentrating up to 1,000 suns (through magnifying lens) onto a smaller sized photovoltaic cell. However, such concentrated solar power requires sophisticated heat sink designs, otherwise the photovoltaic cell overheats, which reduces its efficiency and life. To further exacerbate the concentrated cooling design, the heat sink must be passive, otherwise the power required for active cooling would reduce the overall efficiency and economy.

Crystalline silicon solar cell prices have fallen from $76.67/Watt in 1977 to an estimated $0.74/Watt in 2013. This is seen as evidence supporting Swanson's law, an observation similar to the famous Moore's Law that states that solar cell prices fall 20% for every doubling of industry capacity.

As of 2011, the price of PV modules has fallen by 60% since the summer of 2008, according to Bloomberg New Energy Finance estimates, putting solar power for the first time on a competitive footing with the retail price of electricity in a number of sunny countries; an alternative and consistent price decline figure of 75% from 2007 to 2012 has also been published, though it is unclear whether these figures are specific to the United States or generally global. The levelised cost of electricity (LCOE) from PV is competitive with conventional electricity sources in an expanding list of geographic regions, particularly when the time of generation is included, as electricity is worth more during the day than at night. There has been fierce competition in the supply chain, and further improvements in the levelised cost of energy for solar lie ahead, posing a growing threat to the dominance of fossil fuel generation sources in the next few years. As time progresses, renewable energy technologies generally get cheaper, while fossil fuels generally get more expensive:

The less solar power costs, the more favorably it compares to conventional power, and the more attractive it becomes to utilities and energy users around the globe. Utility-scale solar power can now be delivered in California at prices well below $100/MWh ($0.10/kWh) less than most other peak generators, even those running on low-cost natural gas. Lower solar module costs also stimulate demand from consumer markets where the cost of solar compares very favorably to retail electric rates.

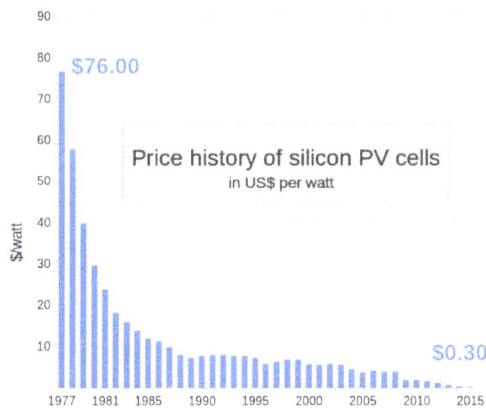

Price per watt history for conventional (c-Si) solar cells since 1977.

As of 2011, the cost of PV has fallen well below that of nuclear power and is set to fall further. The average retail price of solar cells as monitored by the Solarbuzz group fell from $3.50/watt to $2.43/watt over the course of 2011.

For large-scale installations, prices below $1.00/watt were achieved. A module price of 0.60 Euro/watt ($0.78/watt) was published for a large scale 5-year deal in April 2012.

By the end of 2012, the "best in class" module price had dropped to $0.50/watt, and was expected to drop to $0.36/watt by 2017.

In many locations, PV has reached grid parity, which is usually defined as PV production costs at or below retail electricity prices (though often still above the power station prices for coal or gas-fired generation without their distribution and other costs). However, in many countries there is still a need for more access to capital to develop PV projects. To solve this problem securitization has been proposed and used to accelerate development of solar photovoltaic projects. For example, SolarCity offered, the first U.S. asset-backed security in the solar industry in 2013.

Photovoltaic power is also generated during a time of day that is close to peak demand (precedes it) in electricity systems with high use of air conditioning. More generally, it is now evident that, given a carbon price of $50/ton, which would raise the price of coal-fired power by 5c/kWh, solar PV will be cost-competitive in most locations. The declining price of PV has been reflected in rapidly growing installations, totaling about 23 GW in 2011. Although some consolidation is likely in

2012, due to support cuts in the large markets of Germany and Italy, strong growth seems likely to continue for the rest of the decade. Already, by one estimate, total investment in renewables for 2011 exceeded investment in carbon-based electricity generation.

In the case of self consumption payback time is calculated based on how much electricity is not brought from the grid. Additionally, using PV solar power to charge DC batteries, as used in Plug-in Hybrid Electric Vehicles and Electric Vehicles, leads to greater efficiencies. Traditionally, DC generated electricity from solar PV must be converted to AC for buildings, at an average 10% loss during the conversion. An additional efficiency loss occurs in the transition back to DC for battery driven devices and vehicles, and using various interest rates and energy price changes were calculated to find present values that range from $2,057 to $8,213 (analysis from 2009).

For example, in Germany with electricity prices of 0.25 euro/kWh and Insolation of 900 kWh/kW one kW_p will save 225 euro per year and with installation cost of 1700 euro/kW_p means that the system will pay back in less than 7 years.

Manufacturing

Overall the manufacturing process of creating solar photovoltaics is simple in that it does not require the culmination of many complex or moving parts. Because of the solid state nature of PV systems they often have relatively long lifetimes, anywhere from 10 to 30 years. In order to increase electrical output of a PV system the manufacturer must simply add more photovoltaic components and because of this economies of scale are important for manufacturers as costs decrease with increasing output.

While there are many types of PV systems known to be effective, crystalline silicon PV accounted for around 90% of the worldwide production of PV in 2013. Manufacturing silicon PV systems has several steps. First, polysilicon is processed from mined quartz until it is very pure (semi-conductor grade). This is melted down when small amounts of Boron, a group III element, are added to make a p-type semiconductor rich in electron holes. Typically using a seed crystal, an ingot of this solution is grown from the liquid polycrystalline. The ingot may also be cast in a mold. Wafers of this semiconductor material are cut from the bulk material with wire saws, and then go through surface etching before being cleaned. Next, the wafers are placed into a phosphorus vapor deposition furnace which lays a very thin layer of phosphorus, a group V element, which creates an N-type semiconducting surface. To reduce energy losses an anti-reflective coating is added to the surface, along with electrical contacts. After finishing the cell, cells are connected via electrical circuit according to the specific application and prepared for shipping and installation.

Crystalline silicon photovoltaics are only one type of PV, and while they represent the majority of solar cells produced currently there are many new and promising technologies that have the potential to be scaled up to meet future energy needs.

Another newer technology, thin-film PV, are manufactured by depositing semiconducting layers on substrate in vacuum. The substrate is often glass or stainless-steel, and these semiconducting layers are made of many types of materials including cadmium telluride (CdTe), copper indium diselenide (CIS), copper indium gallium diselenide (CIGS), and amorphous silicon (a-Si). After being deposited onto the substrate the semiconducting layers are separated and connected by electrical circuit by laser-scribing. Thin-film photovoltaics now make up around 20% of the overall production of PV because of the reduced materials requirements and cost to manufacture modules consisting of thin-films as compared to silicon-based wafers.

Other emerging PV technologies include organic, dye-sensitized, quantum-dot, and Perovskite photovoltaics. OPVs fall into the thin-film category of manufacturing, and typically operate around the 12% efficiency range which is lower than the 12–21% typically seen by silicon based PVs. Because organic photovoltaics require very high purity and are relatively reactive they must be encapsulated which vastly increases cost of manufacturing and meaning that they are not feasible for large scale up. Dye-sensitized PVs are similar in efficiency to OPVs but are significantly easier to manufacture. However these dye-sensitized photovoltaics present storage problems because the liquid electrolyte is toxic and can potentially permeate the plastics used in the cell. Quantum dot solar cells are quantum dot sensitized DSSCs and are solution processed meaning they are potentially scalable, but currently they peak at 12% efficiency. Perovskite solar cells are a very efficient solar energy converter and have excellent optoelectric properties for photovoltaic purposes, but they are expensive and difficult to manufacture.

Applications

A photovoltaic system, or solar PV system is a power system designed to supply usable solar power by means of photovoltaics. It consists of an arrangement of several components, including solar panels to absorb and directly convert sunlight into electricity, a solar inverter to change the electric current from DC to AC, as well as mounting, cabling and other electrical accessories. PV systems range from small, roof-top mounted or building-integrated systems with capacities from a few to several tens of kilowatts, to large utility-scale power stations of hundreds of megawatts. Nowadays, most PV systems are grid-connected, while stand-alone systems only account for a small portion of the market.

- Rooftop and building integrated systems

Rooftop PV on half-timbered house

Photovoltaic arrays are often associated with buildings: either integrated into them, mounted on them or mounted nearby on the ground. Rooftop PV systems are most often retrofitted into existing buildings, usually mounted on top of the existing roof structure or on the existing walls. Alternatively, an array can be located separately from the building but connected by cable to supply power for the building. Building-integrated photovoltaics

(BIPV) are increasingly incorporated into the roof or walls of new domestic and industrial buildings as a principal or ancillary source of electrical power. Roof tiles with integrated PV cells are sometimes used as well. Provided there is an open gap in which air can circulate, rooftop mounted solar panels can provide a passive cooling effect on buildings during the day and also keep accumulated heat in at night. Typically, residential rooftop systems have small capacities of around 5–10 kW, while commercial rooftop systems often amount to several hundreds of kilowatts. Although rooftop systems are much smaller than ground-mounted utility-scale power plants, they account for most of the worldwide installed capacity.

- Concentrator photovoltaics

Concentrator photovoltaics (CPV) is a photovoltaic technology that contrary to conventional flat-plate PV systems uses lenses and curved mirrors to focus sunlight onto small, but highly efficient, multi-junction (MJ) solar cells. In addition, CPV systems often use solar trackers and sometimes a cooling system to further increase their efficiency. Ongoing research and development is rapidly improving their competitiveness in the utility-scale segment and in areas of high solar insolation.

- Photovoltaic thermal hybrid solar collector

Photovoltaic thermal hybrid solar collector (PVT) are systems that convert solar radiation into thermal and electrical energy. These systems combine a solar PV cell, which converts sunlight into electricity, with a solar thermal collector, which captures the remaining energy and removes waste heat from the PV module. The capture of both electricity and heat allow these devices to have higher exergy and thus be more overall energy efficient than solar PV or solar thermal alone.

- Power stations

Many utility-scale solar farms have been constructed all over the world. As of 2015, the 579-megawatt (MW_{AC}) Solar Star is the world's largest photovoltaic power station, followed by the Desert Sunlight Solar Farm and the Topaz Solar Farm, both with a capacity of 550 MW_{AC}, constructed by US-company First Solar, using CdTe modules, a thin-film PV technology. All three power stations are located in the Californian desert. Many solar farms around the world are integrated with agriculture and some use innovative solar tracking systems that follow the sun's daily path across the sky to generate more electricity than conventional fixed-mounted systems. There are no fuel costs or emissions during operation of the power stations.

- Rural electrification

Developing countries where many villages are often more than five kilometers away from grid power are increasingly using photovoltaics. In remote locations in India a rural lighting program has been providing solar powered LED lighting to replace kerosene lamps. The solar powered lamps were sold at about the cost of a few months' supply of kerosene. Cuba is working to provide solar power for areas that are off grid. More complex applications of off-grid solar energy use include 3D printers. RepRap 3D printers have been solar powered with photovoltaic technology, which enables distributed manufacturing for sustainable development. These are areas where the social costs and benefits offer an excellent case for going solar, though the lack of profitability has relegated such endeavors to humanitarian efforts. However, in 1995 solar rural electrification projects had been found to be difficult to sustain due to unfavorable economics, lack of technical support, and a legacy of ulterior motives of north-to-south technology transfer.

- Standalone systems

Until a decade or so ago, PV was used frequently to power calculators and novelty devices. Improvements in integrated circuits and low power liquid crystal displays make it possible to power such devices for several years between battery changes, making PV use less common. In contrast, solar powered remote fixed devices have seen increasing use recently in locations where significant connection cost makes grid power prohibitively expensive. Such applications include solar lamps, water pumps, parking meters, emergency telephones, trash compactors, temporary traffic signs, charging stations, and remote guard posts and signals.

- Floatovoltaics

In May 2008, the Far Niente Winery in Oakville, CA pioneered the world's first "floatovoltaic" system by installing 994 photovoltaic solar panels onto 130 pontoons and floating them on the winery's irrigation pond. The floating system generates about 477 kW of peak output and when combined with an array of cells located adjacent to the pond is able to fully offset the winery's electricity consumption. The primary benefit of a floatovoltaic system is that it avoids the need to sacrifice valuable land area that could be used for another purpose. In the case of the Far Niente Winery, the floating system saved three-quarters of an acre that would have been required for a land-based system. That land area can instead be used for agriculture. Another benefit of a floatovoltaic system is that the panels are kept at a lower temperature than they would be on land, leading to a higher efficiency of solar energy conversion. The floating panels also reduce the amount of water lost through evaporation and inhibit the growth of algae.

- In transport

PV has traditionally been used for electric power in space. PV is rarely used to provide motive power in transport applications, but is being used increasingly to provide auxiliary power in boats and cars. Some automobiles are fitted with solar-powered air conditioning to limit interior temperatures on hot days. A self-contained solar vehicle would have limited power and utility, but a solar-charged electric vehicle allows use of solar power for transportation. Solar-powered cars, boats and airplanes have been demonstrated, with the most practical and likely of these being solar cars. The Swiss solar aircraft, Solar Impulse 2, achieved the longest non-stop solo flight in history and plan to make the first solar-powered aerial circumnavigation of the globe in 2015.

- Telecommunication and signaling

Solar PV power is ideally suited for telecommunication applications such as local telephone exchange, radio and TV broadcasting, microwave and other forms of electronic communication links. This is because, in most telecommunication application, storage batteries are already in use and the electrical system is basically DC. In hilly and mountainous terrain, radio and TV signals may not reach as they get blocked or reflected back due to undulating terrain. At these locations, low power transmitters (LPT) are installed to receive and re-transmit the signal for local population.

- Spacecraft applications

Part of Juno's solar array

Solar panels on spacecraft are usually the sole source of power to run the sensors, active heating and cooling, and communications. A battery stores this energy for use when the solar panels are in shadow. In some, the power is also used for spacecraft propulsion—electric propulsion. Spacecraft were one of the earliest applications of photovoltaics, starting with the silicon solar cells used on the Vanguard 1 satellite, launched by the US in 1958. Since then, solar power has been used on missions ranging from the MESSENGER probe to Mercury, to as far out in the solar system as the Juno probe to Jupiter. The largest solar power system flown in space is the electrical system of the International Space Station. To increase the power generated per kilogram, typical spacecraft solar panels use high-cost, high-efficiency, and close-packed rectangular multi-junction solar cells made of gallium arsenide (GaAs) and other semiconductor materials.

- Specialty Power Systems

Photovoltaics may also be incorporated as energy conversion devices for objects at elevated temperatures and with preferable radiative emissivities such as heterogeneous combustors.

Advantages

The 122 PW of sunlight reaching the Earth's surface is plentiful—almost 10,000 times more than the 13 TW equivalent of average power consumed in 2005 by humans. This abundance leads to the suggestion that it will not be long before solar energy will become the world's primary energy source. Additionally, solar electric generation has the highest power density (global mean of 170 W/m²) among renewable energies.

Solar power is pollution-free during use. Production end-wastes and emissions are manageable using existing pollution controls. End-of-use recycling technologies are under development and policies are being produced that encourage recycling from producers.

PV installations can operate for 100 years or even more with little maintenance or intervention after their initial set-up, so after the initial capital cost of building any solar power plant, operating costs are extremely low compared to existing power technologies.

Grid-connected solar electricity can be used locally thus reducing transmission/distribution losses (transmission losses in the US were approximately 7.2% in 1995).

Compared to fossil and nuclear energy sources, very little research money has been invested in the development of solar cells, so there is considerable room for improvement. Nevertheless, experimental high efficiency solar cells already have efficiencies of over 40% in case of concentrating photovoltaic cells and efficiencies are rapidly rising while mass-production costs are rapidly falling.

In some states of the United States, much of the investment in a home-mounted system may be lost if the home-owner moves and the buyer puts less value on the system than the seller. The city of Berkeley developed an innovative financing method to remove this limitation, by adding a tax assessment that is transferred with the home to pay for the solar panels. Now known as PACE, Property Assessed Clean Energy, 30 U.S. states have duplicated this solution.

There is evidence, at least in California, that the presence of a home-mounted solar system can actually increase the value of a home. According to a paper published in April 2011 by the Ernest Orlando Lawrence Berkeley National Laboratory titled An Analysis of the Effects of Residential Photovoltaic Energy Systems on Home Sales Prices in California:

The research finds strong evidence that homes with PV systems in California have sold for a premium over comparable homes without PV systems. More specifically, estimates for average PV premiums range from approximately $3.9 to $6.4 per installed watt (DC) among a large number of different model specifications, with most models coalescing near $5.5/watt. That value corresponds to a premium of approximately $17,000 for a relatively new 3,100 watt PV system (the average size of PV systems in the study).

Photovoltaic System

A photovoltaic system, also PV system or solar power system, is a power system designed to supply usable solar power by means of photovoltaics. It consists of an arrangement of several components, including solar panels to absorb and convert sunlight into electricity, a solar inverter to change the electric current from DC to AC, as well as mounting, cabling and other electrical accessories to set up a working system. It may also use a solar tracking system to improve the system's overall performance and include an integrated battery solution, as prices for storage devices are expected to decline. Strictly speaking, a solar array only encompasses the ensemble of solar panels, the visible part of the PV system, and does not include all the other hardware, often summarized as balance of system (BOS). Moreover, PV systems convert light directly into electricity and shouldn't be confused with other technologies, such as concentrated solar power or solar thermal, used for heating and cooling.

PV systems range from small, rooftop-mounted or building-integrated systems with capacities from a few to several tens of kilowatts, to large utility-scale power stations of hundreds of megawatts. Nowadays, most PV systems are grid-connected, while off-grid or stand-alone systems only account for a small portion of the market.

Operating silently and without any moving parts or environmental emissions, PV systems have developed from being niche market applications into a mature technology used for mainstream electricity generation. A rooftop system recoups the invested energy for its manufacturing and installation within 0.7 to 2 years and produces about 95 percent of net clean renewable energy over a 30-year service lifetime.

Due to the exponential growth of photovoltaics, prices for PV systems have rapidly declined in recent years. However, they vary by market and the size of the system. In 2014, prices for residential 5-kilowatt systems in the United States were around $3.29 per watt, while in the highly penetrated German market, prices for rooftop systems of up to 100 kW declined to €1.24 per watt. Nowadays, solar PV modules account for less than half of the system's overall cost, leaving the rest to the remaining BOS-components and to soft costs, which include customer acquisition, permitting, inspection and interconnection, installation labor and financing costs.

Modern System

Overview

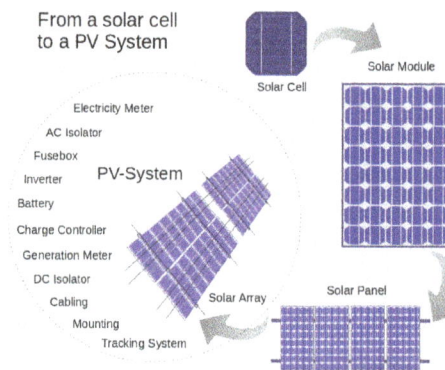

From a solar cell
to a PV System

Solar Module

Solar Cell

Electricity Meter
AC Isolator
Fusebox
Inverter   PV-System
Battery
Charge Controller
Generation Meter
DC Isolator
Cabling   Solar Array   Solar Panel
Mounting
Tracking System

Diagram of the possible components of a photovoltaic system

A photovoltaic system converts the sun's radiation into usable electricity. It comprises the solar array and the balance of system components. PV systems can be categorized by various aspects, such as, grid-connected vs. stand alone systems, building-integrated vs. rack-mounted systems, residential vs. utility systems, distributed vs. centralized systems, rooftop vs. ground-mounted systems, tracking vs. fixed-tilt systems, and new constructed vs. retrofitted systems. Other distinctions may include, systems with microinverters vs. central inverter, systems using crystalline silicon vs. thin-film technology, and systems with modules from Chinese vs. European and U.S.-manufacturers.

About 99 percent of all European and 90 percent of all U.S. solar power systems are connected to the electrical grid, while off-grid systems are somewhat more common in Australia and South Korea. PV systems rarely use battery storage. This may change soon, as government incentives for distributed energy storage are being implemented and investments in storage solutions are gradually becoming economically viable for small systems. A solar array of a typical residential

PV system is rack-mounted on the roof, rather than integrated into the roof or facade of the building, as this is significantly more expensive. Utility-scale solar power stations are ground-mounted, with fixed tilted solar panels rather than using expensive tracking devices. Crystalline silicon is the predominant material used in 90 percent of worldwide produced solar modules, while rival thin-film has lost market-share in recent years. About 70 percent of all solar cells and modules are produced in China and Taiwan, leaving only 5 percent to European and US-manufacturers. The installed capacity for both, small rooftop systems and large solar power stations is growing rapidly and in equal parts, although there is a notable trend towards utility-scale systems, as the focus on new installations is shifting away from Europe to sunnier regions, such as the Sunbelt in the U.S., which are less opposed to ground-mounted solar farms and cost-effectiveness is more emphasized by investors.

Driven by advances in technology and increases in manufacturing scale and sophistication, the cost of photovoltaics is declining continuously. There are several million PV systems distributed all over the world, mostly in Europe, with 1.4 million systems in Germany alone– as well as North America with 440,000 systems in the United States, The energy conversion efficiency of a conventional solar module increased from 15 to 20 percent over the last 10 years and a PV system recoups the energy needed for its manufacture in about 2 years. In exceptionally irradiated locations, or when thin-film technology is used, the so-called energy payback time decreases to one year or less. Net metering and financial incentives, such as preferential feed-in tariffs for solar-generated electricity, have also greatly supported installations of PV systems in many countries. The levelised cost of electricity from large-scale PV systems has become competitive with conventional electricity sources in an expanding list of geographic regions, and grid parity has been achieved in about 30 different countries.

As of 2015, the fast-growing global PV market is rapidly approaching the 200 GW mark – about 40 times the installed capacity of 2006. Photovoltaic systems currently contribute about 1 percent to worldwide electricity generation. Top installers of PV systems in terms of capacity are currently China, Japan and the United States, while half of the world's capacity is installed in Europe, with Germany and Italy supplying 7% to 8% of their respective domestic electricity consumption with solar PV. The International Energy Agency expects solar power to become the world's largest source of electricity by 2050, with solar photovoltaics and concentrated solar thermal contributing 16% and 11% to the global demand, respectively.

Grid-connection

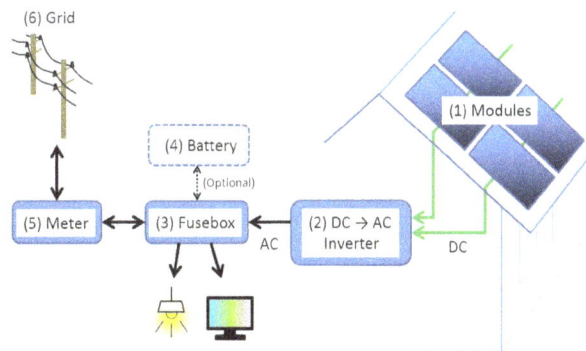

Schematics of a typical residential PV system

A grid connected system is connected to a larger independent grid (typically the public electricity grid) and feeds energy directly into the grid. This energy may be shared by a residential or commercial building before or after the revenue measurement point. The difference being whether the credited energy production is calculated independently of the customer's energy consumption (feed-in tariff) or only on the difference of energy (net metering). Grid connected systems vary in size from residential (2–10 kW_p) to solar power stations (up to 10s of MW_p). This is a form of decentralized electricity generation. The feeding of electricity into the grid requires the transformation of DC into AC by a special, synchronising grid-tie inverter. In kilowatt-sized installations the DC side system voltage is as high as permitted (typically 1000V except US residential 600 V) to limit ohmic losses. Most modules (60 or 72 crystalline silicon cells) generate 160 W to 300 W at 36 volts. It is sometimes necessary or desirable to connect the modules partially in parallel rather than all in series. One set of modules connected in series is known as a 'string'.

Scale of System

Photovoltaic systems are generally categorized into three distinct market segments: residential rooftop, commercial rooftop, and ground-mount utility-scale systems. Their capacities range from a few kilowatts to hundreds of megawatts. A typical residential system is around 10 kilowatts and mounted on a sloped roof, while commercial systems may reach a megawatt-scale and are generally installed on low-slope or even flat roofs. Although rooftop mounted systems are small and display a higher cost per watt than large utility-scale installations, they account for the largest share in the market. There is, however, a growing trend towards bigger utility-scale power plants, especially in the "sunbelt" region of the planet.

Utility-scale

Perovo Solar Park in Ukraine

Large utility-scale solar parks or farms are power stations and capable of providing an energy supply to large numbers of consumers. Generated electricity is fed into the transmission grid powered by central generation plants (grid-connected or grid-tied plant), or combined with one, or many, domestic electricity generators to feed into a small electrical grid (hybrid plant). In rare cases generated electricity is stored or used directly by island/standalone plant. PV systems are generally designed in order to ensure the highest energy yield for a given investment. Some large photovoltaic power stations such as Solar Star, Waldpolenz Solar Park and Topaz Solar Farm cover tens or hundreds of hectares and have power outputs up to hundreds of megawatts.

Rooftop, Mobile, and Portable

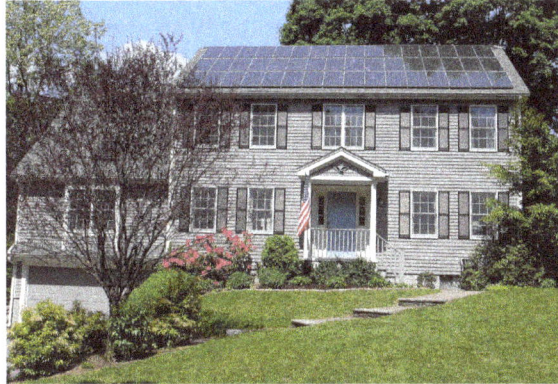
Rooftop system near Boston, USA

A small PV system is capable of providing enough AC electricity to power a single home, or even an isolated device in the form of AC or DC electric. For example, military and civilian Earth observation satellites, street lights, construction and traffic signs, electric cars, solar-powered tents, and electric aircraft may contain integrated photovoltaic systems to provide a primary or auxiliary power source in the form of AC or DC power, depending on the design and power demands. In 2013, rooftop systems accounted for 60 percent of worldwide installations. However, there is a trend away from rooftop and towards utility-scale PV systems, as the focus of new PV installations is also shifting from Europe to countries in the sunbelt region of the planet where opposition to ground-mounted solar farms is less accentuated.

Portable and mobile PV systems provide electrical power independent of utility connections, for "off the grid" operation. Such systems are so commonly used on recreational vehicles and boats that there are retailers specializing in these applications and products specifically targeted to them. Since recreational vehicles (RV) normally carry batteries and operate lighting and other systems on nominally 12-volt DC power, RV PV systems normally operate in a voltage range chosen to charge 12-volt batteries directly, and addition of a PV system requires only panels, a charge controller, and wiring.

Building-integrated

BAPV wall near Barcelona, Spain

In urban and suburban areas, photovoltaic arrays are commonly used on rooftops to supplement power use; often the building will have a connection to the power grid, in which case the energy produced by the PV array can be sold back to the utility in some sort of net metering agreement. Some utilities, such as Solvay Electric in Solvay, NY, use the rooftops of commercial customers and telephone poles to support their use of PV panels. Solar trees are arrays that, as the name implies, mimic the look of trees, provide shade, and at night can function as street lights.

Performance

Uncertainties in revenue over time relate mostly to the evaluation of the solar resource and to the performance of the system itself. In the best of cases, uncertainties are typically 4% for year-to-year climate variability, 5% for solar resource estimation (in a horizontal plane), 3% for estimation of irradiation in the plane of the array, 3% for power rating of modules, 2% for losses due to dirt and soiling, 1.5% for losses due to snow, and 5% for other sources of error. Identifying and reacting to manageable losses is critical for revenue and O&M efficiency. Monitoring of array performance may be part of contractual agreements between the array owner, the builder, and the utility purchasing the energy produced. Recently, a method to create "synthetic days" using readily available weather data and verification using the Open Solar Outdoors Test Field make it possible to predict photovoltaic systems performance with high degrees of accuracy. This method can be used to then determine loss mechanisms on a local scale - such as those from snow or the effects of surface coatings (e.g. hydrophobic or hydrophilic) on soiling or snow losses. (Although in heavy snow environments with severe ground interference can result in annual losses from snow of 30%.) Access to the Internet has allowed a further improvement in energy monitoring and communication. Dedicated systems are available from a number of vendors. For solar PV systems that use microinverters (panel-level DC to AC conversion), module power data is automatically provided. Some systems allow setting performance alerts that trigger phone/email/text warnings when limits are reached. These solutions provide data for the system owner and the installer. Installers are able to remotely monitor multiple installations, and see at-a-glance the status of their entire installed base.

Components

Balance of System

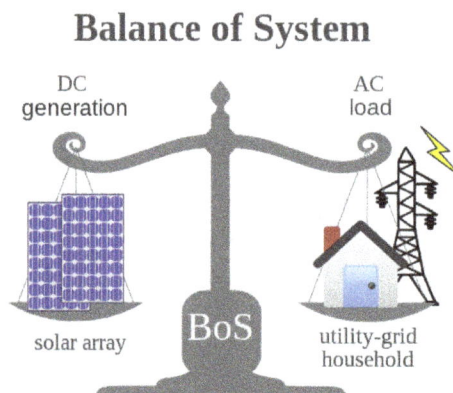

The balance of system components of a PV system (BOS) balance the power-generating subsystem of the solar array (left side) with the power-using side of the AC-household devices and the utility grid (right side).

A photovoltaic system for residential, commercial, or industrial energy supply consists of the solar array and a number of components often summarized as the balance of system (BOS). This term is synonymous with "Balance of Plant" q.v. at. BOS-components include power-conditioning equipment and structures for mounting, typically one or more DC to AC power converters, also known as inverters, an energy storage device, a racking system that supports the solar array, electrical wiring and interconnections, and mounting for other components.

Optionally, a balance of system may include any or all of the following: renewable energy credit revenue-grade meter, maximum power point tracker (MPPT), battery system and charger, GPS solar tracker, energy management software, solar irradiance sensors, anemometer, or task-specific accessories designed to meet specialized requirements for a system owner. In addition, a CPV system requires optical lenses or mirrors and sometimes a cooling system.

The terms *"solar array"* and *"PV system"* are often incorrectly used interchangeably, despite the fact that the solar array does not encompass the entire system. Moreover, *"solar panel"* is often used as a synonym for *"solar module"*, although a panel consists of a string of several modules. The term *"solar system"* is also an often used misnomer for a PV system.

Solar Array

Conventional c-Si solar cells, normally wired in series, are encapsulated in a solar module to protect them from the weather. The module consists of a tempered glass as cover, a soft and flexible encapsulant, a rear backsheet made of a weathering and fire-resistant material and an aluminium frame around the outer edge. Electrically connected and mounted on a supporting structure, solar modules build a string of modules, often called solar panel. A solar array consists of one or many such panels. A photovoltaic array, or solar array, is a linked collection of solar panels. The power that one module can produce is seldom enough to meet requirements of a home or a business, so the modules are linked together to form an array. Most PV arrays use an inverter to convert the DC power produced by the modules into alternating current that can power lights, motors, and other loads. The modules in a PV array are usually first connected in series to obtain the desired voltage; the individual strings are then connected in parallel to allow the system to produce more current. Solar panels are typically measured under STC (standard test conditions) or PTC (PVUSA test conditions), in watts. Typical panel ratings range from less than 100 watts to over 400 watts. The array rating consists of a summation of the panel ratings, in watts, kilowatts, or megawatts.

Module and Efficiency

A typical "150 watt" PV module is about a square meter in size. Such a module may be expected to produce 0.75 kilowatt-hour (kWh) every day, on average, after taking into account the weather and the latitude, for an insolation of 5 sun hours/day. In the last 10 years, the efficiency of average commercial wafer-based crystalline silicon modules increased from about 12% to 16% and CdTe module efficiency increased from 9% to 13% during same period. Module output and life degraded by increased temperature. Allowing ambient air to flow over, and if possible behind, PV modules reduces this problem. Effective module lives are typically 25 years or more. The payback period for an investment in a PV solar installation varies greatly and is typically less useful than a calculation of return on investment. While it is typically calculated to be between 10 and 20 years, the financial payback period can be far shorter with incentives.

Fixed tilt solar array in of crystalline silicon panels in Canterbury, New Hampshire, United States.

Solar array of a solar farm with a few thousand solar modules on the island of Majorca, Spain.

Due to the low voltage of an individual solar cell (typically ca. 0.5V), several cells are in series in the manufacture of a "laminate". The laminate is assembled into a protective weatherproof enclosure, thus making a photovoltaic module or solar panel. Modules may then be strung together into a photovoltaic array. In 2012, solar panels available for consumers can have an efficiency of up to about 17%, while commercially available panels can go as far as 27%. It has been recorded that a group from The Fraunhofer Institute for Solar Energy Systems have created a cell that can reach 44.7% efficiency, which makes scientists' hopes of reaching the 50% efficiency threshold a lot more feasible.

Shading and Dirt

Photovoltaic cell electrical output is extremely sensitive to shading. The effects of this shading are well known. When even a small portion of a cell, module, or array is shaded, while the remainder is in sunlight, the output falls dramatically due to internal 'short-circuiting' (the electrons reversing course through the shaded portion of the p-n junction). If the current drawn from the series string of cells is no greater than the current that can be produced by the shaded cell, the current (and so power) developed by the string is limited. If enough voltage is available from the rest of the cells in a string, current will be forced through the cell by breaking down the junction in the shaded portion. This breakdown voltage in common cells is between 10 and 30 volts. Instead of adding to the power produced by the panel, the shaded cell absorbs power, turning it into heat. Since the reverse voltage of a shaded cell is much greater than the forward voltage of an illuminated cell, one shaded cell can absorb the power of many other cells in the string, disproportionately affecting panel output. For example, a shaded cell may drop 8 volts, instead of adding 0.5 volts, at a particular current level, thereby absorbing the power produced by 16 other cells. It is, thus important that a PV installation is not shaded by trees or other obstructions.

Several methods have been developed to determine shading losses from trees to PV systems over both large regions using LiDAR, but also at an individual system level using sketchup. Most modules have bypass diodes between each cell or string of cells that minimize the effects of shading and only lose the power of the shaded portion of the array. The main job of the bypass diode is to eliminate hot spots that form on cells that can cause further damage to the array, and cause fires. Sunlight can be absorbed by dust, snow, or other impurities at the surface of the module. This can reduce the light that strikes the cells. In general these losses aggregated over the year are small even for locations in Canada. Maintaining a clean module surface will increase output performance over the life of the module. Google found that cleaning the flat mounted solar panels after 15 months increased their output by almost 100%, but that the 5% tilted arrays were adequately cleaned by rainwater.

Insolation and Energy

Solar insolation is made up of direct, diffuse, and reflected radiation. The absorption factor of a PV cell is defined as the fraction of incident solar irradiance that is absorbed by the cell. At high noon on a cloudless day at the equator, the power of the sun is about 1 kW/m², on the Earth's surface, to a plane that is perpendicular to the sun's rays. As such, PV arrays can track the sun through each day to greatly enhance energy collection. However, tracking devices add cost, and require maintenance, so it is more common for PV arrays to have fixed mounts that tilt the array and face solar noon (approximately due south in the Northern Hemisphere or due north in the Southern Hemisphere). The tilt angle, from horizontal, can be varied for season, but if fixed, should be set to give optimal array output during the peak electrical demand portion of a typical year for a stand-alone system. This optimal module tilt angle is not necessarily identical to the tilt angle for maximum annual array energy output. The optimization of the photovoltaic system for a specific environment can be complicated as issues of solar flux, soiling, and snow losses should be taken into effect. In addition, recent work has shown that spectral effects can play a role in optimal photovoltaic material selection. For example, the spectral albedo can play a significant role in output depending on the surface around the photovoltaic system and the type of solar cell material. For the weather and latitudes of the United States and Europe, typical insolation ranges from 4 kWh/m²/day in northern climes to 6.5 kWh/m²/day in the sunniest regions. A photovoltaic installation in the southern latitudes of Europe or the United States may expect to produce 1 kWh/m²/day. A typical 1 kW photovoltaic installation in Australia or the southern latitudes of Europe or United States, may produce 3.5–5 kWh per day, dependent on location, orientation, tilt, insolation and other factors. In the Sahara desert, with less cloud cover and a better solar angle, one could ideally obtain closer to 8.3 kWh/m²/day provided the nearly ever present wind would not blow sand onto the units. The area of the Sahara desert is over 9 million km². 90,600 km², or about 1%, could generate as much electricity as all of the world's power plants combined.

Mounting

Modules are assembled into arrays on some kind of mounting system, which may be classified as ground mount, roof mount or pole mount. For solar parks a large rack is mounted on the ground, and the modules mounted on the rack. For buildings, many different racks have been devised for pitched roofs. For flat roofs, racks, bins and building integrated solutions are used. Solar panel racks mounted on top of poles can be stationary or moving. Side-of-pole mounts are suitable for situations where a pole has something else mounted at its top, such as a light fixture or an antenna. Pole mounting raises what would otherwise be a ground mounted array above weed shadows and livestock, and may satisfy electrical code requirements regarding inaccessibility of exposed wiring. Pole mounted panels are open to more cooling air on their underside, which increases performance. A multiplicity of pole top racks can be formed into a parking carport or other shade structure. A rack which does not follow the sun from left to right may allow seasonal adjustment up or down.

A 23-year-old, ground mounted PV system from the 1980s on a North Frisian Island, Germany. The modules conversion efficiency was only 12%.

Cabling

Due to their outdoor usage, solar cables are specifically designed to be resistant against UV radiation and extremely high temperature fluctuations and are generally unaffected by the weather. A number of standards specify the usage of electrical wiring in PV systems, such as the IEC 60364 by the International Electrotechnical Commission, in section 712 *"Solar photovoltaic (PV) power supply systems"*, the British Standard BS 7671, incorporating regulations relating to microgeneration and photovoltaic systems, and the US UL4703 standard, in subject 4703 *"Photovoltaic Wire"*.

Tracker

A 1998 model of a passive solar tracker, viewed from underneath.

A solar tracking system tilts a solar panel throughout the day. Depending on the type of tracking system, the panel is either aimed directly at the sun or the brightest area of a partly clouded sky. Trackers greatly enhance early morning and late afternoon performance, increasing the total amount of power produced by a system by about 20–25% for a single axis tracker and about 30% or more for a dual axis tracker, depending on latitude. Trackers are effective in regions that receive a large portion of sunlight directly. In diffuse light (i.e. under cloud or fog), tracking has little or no value. Because most concentrated photovoltaics systems are very sensitive to the sunlight's angle, tracking systems allow them to produce useful power for more than a brief period each day. Tracking systems improve performance for two main reasons. First, when a solar panel is perpendicular to the sunlight, it receives more light on its surface than if it were angled. Second, direct light is used more efficiently than angled light. Special Anti-reflective coatings can improve solar panel efficiency for direct and angled light, somewhat reducing the benefit of tracking.

Trackers and sensors to optimise the performance are often seen as optional, but tracking systems can increase viable output by up to 45%. PV arrays that approach or exceed one megawatt often use solar trackers. Accounting for clouds, and the fact that most of the world is not on the equator, and that the sun sets in the evening, the correct measure of solar power is insolation – the average number of kilowatt-hours per square meter per day. For the weather and latitudes of the United States and Europe, typical insolation ranges from 2.26 kWh/m²/day in northern climes to 5.61 kWh/m²/day in the sunniest regions.

For large systems, the energy gained by using tracking systems can outweigh the added complexity (trackers can increase efficiency by 30% or more). For very large systems, the added maintenance of tracking is a substantial detriment. Tracking is not required for flat panel and low-concentration photovoltaic systems. For high-concentration photovoltaic systems, dual axis tracking is a necessity. Pricing trends affect the balance between adding more stationary solar panels versus having fewer panels that track. When solar panel prices drop, trackers become a less attractive option.

Inverter

Central inverter with AC and DC disconnects (on the side), monitoring gateway, transformer isolation and interactive LCD.

Systems designed to deliver alternating current (AC), such as grid-connected applications need an inverter to convert the direct current (DC) from the solar modules to AC. Grid connected inverters must supply AC electricity in sinusoidal form, synchronized to the grid frequency, limit feed in voltage to no higher than the grid voltage and disconnect from the grid if the grid voltage is turned off. Islanding inverters need only produce regulated voltages and frequencies in a sinusoidal wave-shape as no synchronisation or co-ordination with grid supplies is required.

String inverter (left), generation meter, and AC disconnect (right). A modern 2013 installation in Vermont, United States.

A solar inverter may connect to a string of solar panels. In some installations a solar micro-inverter is connected at each solar panel. For safety reasons a circuit breaker is provided both on the AC and DC side to enable maintenance. AC output may be connected through an electricity meter into the public grid. The number of modules in the system determines the total DC watts capable of being generated by the solar array; however, the inverter ultimately governs the amount of AC watts that can be distributed for consumption. For example, a PV system comprising 11 kilowatts DC (kW_{DC}) worth of PV modules, paired with one 10-kilowatt AC (kW_{AC}) inverter, will be limited to the inverter's output of 10 kW. As of 2014, conversion efficiency for state-of-the-art converters reached more than 98 percent. While string inverters are used in residential to medium-sized commercial PV systems, central inverters cover the large commercial and utility-scale market. Market-share for central and string inverters are about 50 percent and 48 percent, respectively, leaving less than 2 percent to micro-inverters.

Maximum power point tracking (MPPT) is a technique that grid connected inverters use to get the maximum possible power from the photovoltaic array. In order to do so, the inverter's MPPT system digitally samples the solar array's ever changing power output and applies the proper resistance to find the optimal *maximum power point*.

Anti-islanding is a protection mechanism that immediately shuts down the inverter preventing it from generating AC power when the connection to the load no longer exists. This happens, for example, in the case of a blackout. Without this protection, the supply line would become an "island" with power surrounded by a "sea" of unpowered lines, as the solar array continues to deliver DC power during the power outage. Islanding is a hazard to utility workers, who may not realize that an AC circuit is still powered, and it may prevent automatic re-connection of devices.

Inverter/Converter Market in 2014				
Type	**Power**	**Efficiency**	**Market Share**	**Remarks**
String inverter	up to 100 kW$_p$[c]	98%	50%	Cost[b] €0.15 per watt-peak. Easy to replace.
Central inverter	above 100 kW$_p$	98.5%	48%	€0.10 per watt-peak. High reliability. Often sold along with a service contract.
Micro-inverter	module power range	90%–95%	1.5%	€0.40 per watt-peak. Ease of replacement concerns.
DC/DC converter Power optimizer	module power range	98.8%	n.a.	€0.40 per watt-peak. Ease of replacement concerns. Inverter is still needed. About 0.75 GW$_p$ installed in 2013.

Battery

Although still expensive, PV systems increasingly use rechargeable batteries to store a surplus to be later used at night. Batteries used for grid-storage also stabilize the electrical grid by leveling out peak loads, and play an important role in a smart grid, as they can charge during periods of low demand and feed their stored energy into the grid when demand is high.

Common battery technologies used in today's PV systems include the valve regulated lead-acid battery— a modified version of the conventional lead–acid battery, nickel–cadmium and lithium-ion batteries. Compared to the other types, lead-acid batteries have a shorter lifetime and lower energy density. However, due to their high reliability, low self discharge as well as low investment and maintenance costs, they are currently the predominant technology used in small-scale, residential PV systems, as lithium-ion batteries are still being developed and about 3.5 times as expensive as lead-acid batteries. Furthermore, as storage devices for PV systems are stationary, the lower energy and power density and therefore higher weight of lead-acid batteries are not as critical as, for example, in electric transportation. Other rechargeable batteries that are considered for distributed PV systems include sodium–sulfur and vanadium redox batteries, two prominent types of a molten salt and a flow battery, respectively. In 2015, Tesla motors launched the Powerwall, a rechargeable lithium-ion battery with the aim to revolutionize energy consumption.

PV systems with an integrated battery solution also need a charge controller, as the varying voltage and current from the solar array requires constant adjustment to prevent damage from overcharging. Basic charge controllers may simply turn the PV panels on and off, or may meter out pulses of energy as needed, a strategy called PWM or pulse-width modulation. More advanced charge controllers will incorporate MPPT logic into their battery charging algorithms. Charge controllers may also divert energy to some purpose other than battery charging. Rather than simply shut off the free PV energy when not needed, a user may choose to heat air or water once the battery is full.

Monitoring and Metering

The metering must be able to accumulate energy units in both directions or two meters must be used. Many meters accumulate bidirectionally, some systems use two meters, but a unidirectional meter (with detent) will not accumulate energy from any resultant feed into the grid. In some countries, for installations over 30 kW$_p$ a frequency and a voltage monitor with disconnection of all phases is required. This is done where more solar power is being generated than can be accommodated by the utility, and the excess can not either be exported or stored. Grid operators historically have needed to provide transmission lines and generation capacity. Now they need to also provide storage. This is normally hydro-storage, but other means of storage are used. Initially storage was used so that baseload generators could operate at full output. With variable renewable energy, storage is needed to allow power generation whenever it is available, and consumption whenever it is needed.

A Canadian electricity meter

The two variables a grid operator have are storing electricity for *when* it is needed, or transmitting it to *where* it is needed. If both of those fail, installations over 30kWp can automatically shut down, although in practice all inverters maintain voltage regulation and stop supplying power if the load is inadequate. Grid operators have the option of curtailing excess generation from large systems, although this is more commonly done with wind power than solar power, and results in a substantial loss of revenue. Three-phase inverters have the unique option of supplying reactive power which can be advantageous in matching load requirements.

Photovoltaic systems need to be monitored to detect breakdown and optimize their operation. There are several *photovoltaic monitoring* strategies depending on the output of the installation and its nature. Monitoring can be performed on site or remotely. It can measure production only, retrieve all the data from the inverter or retrieve all of the data from the communicating equipment (probes, meters, etc.). Monitoring tools can be dedicated to supervision only or offer additional functions. Individual inverters and battery charge controllers may include monitoring using manufacturer specific protocols and software. Energy metering of an inverter may be of limited accuracy and not suitable for revenue metering purposes. A third-party data acquisition system can monitor multiple inverters, using the inverter manufacturer's protocols, and also acquire weather-related information. Independent smart meters may measure the total energy production of a PV array system. Separate measures such as satellite image analysis or a solar radiation meter (a pyranometer) can be used to estimate total insolation for comparison. Data collected from a monitoring system can be displayed remotely over the World Wide Web, such as OSOTF.

Other Systems

This section includes systems that are either highly specialized and uncommon or still an emerging new technology with limited significance. However, standalone or off-grid systems take a special place. They were the most common type of systems during the 1980s and 1990s, when PV technology was still very expensive and a pure niche market of small scale applications. Only in places where no electrical grid was available, they were economically viable. Although new stand-alone systems are still being deployed all around the world, their contribution to the overall installed photovoltaic capacity is decreasing. In Europe, off-grid systems account for 1 percent of installed capacity. In the United States, they account for about 10 percent. Off-grid systems are still common in Australia and South Korea, and in many developing countries.

CPV

Concentrator photovoltaic (CPV) in Catalonia, Spain

Concentrator photovoltaics (CPV) and *high concentrator photovoltaic* (HCPV) systems use optical lenses or curved mirrors to concentrate sunlight onto small but highly efficient solar cells. Besides concentrating optics, CPV systems sometime use solar trackers and cooling systems and are more expensive.

Especially HCPV systems are best suited in location with high solar irradiance, concentrating sunlight up to 400 times or more, with efficiencies of 24–28 percent, exceeding those of regular systems. Various designs of CPV and HCPV systems are commercially available but not very common. However, ongoing research and development is taking place.

CPV is often confused with CSP (concentrated solar power) that does not use photovoltaics. Both technologies favor locations that receive much sunlight and are directly competing with each other.

Hybrid

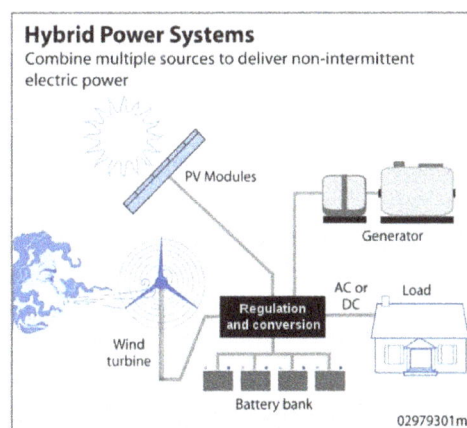

A wind-solar PV hybrid system

A hybrid system combines PV with other forms of generation, usually a diesel generator. Biogas is also used. The other form of generation may be a type able to modulate power output as a function of demand. However more than one renewable form of energy may be used e.g. wind. The photovoltaic power generation serves to reduce the consumption of non renewable fuel. Hybrid systems are most often found on islands. Pellworm island in Germany and Kythnos island in Greece are notable examples (both are combined with wind). The Kythnos plant has reduced diesel consumption by 11.2%.

In 2015, a case-study conducted in seven countries concluded that in all cases generating costs can be reduced by hybridising mini-grids and isolated grids. However, financing costs for such hybrids are crucial and largely depend on the ownership structure of the power plant. While cost reductions for state-owned utilities can be significant, the study also identified economic benefits to be insignificant or even negative for non-public utilities, such as independent power producers.

There has also been recent work showing that the PV penetration limit can be increased by deploying a distributed network of PV+CHP hybrid systems in the U.S. The temporal distribution of solar flux, electrical and heating requirements for representative U.S. single family residences were analyzed and the results clearly show that hybridizing CHP with PV can enable additional PV deployment above what is possible with a conventional centralized electric generation system. This theory was reconfirmed with numerical simulations using per second solar flux data to determine that the necessary battery backup to provide for such a hybrid system is possible with relatively small and inexpensive battery systems. In addition, large PV+CHP systems are possible for institutional buildings, which again provide back up for intermittent PV and reduce CHP runtime.

- PVT system (hybrid PV/T), also known as *photovoltaic thermal hybrid solar collectors* convert solar radiation into thermal and electrical energy. Such a system combines a solar (PV) module with a solar thermal collector in an complementary way.

- CPVT system. A *concentrated photovoltaic thermal hybrid* (CPVT) system is similar to a PVT system. It uses concentrated photovoltaics (CPV) instead of conventional PV technology, and combines it with a solar thermal collector.

- CPV/CSP system. A novel solar CPV/CSP hybrid system has been proposed recently, combining concentrator photovoltaics with the non-PV technology of concentrated solar power (CSP), or also known as concentrated solar thermal.

- PV diesel system. It combines a photovoltaic system with a diesel generator. Combinations with other renewables are possible and include wind turbines.

Floating Solar Arrays

Floating solar arrays are PV systems that float on the surface of drinking water reservoirs, quarry lakes, irrigation canals or remediation and tailing ponds. A small number of such systems exist in France, India, Japan, South Korea, the United Kingdom, Singapore and the United States.

The systems are said to have advantages over photovoltaics on land. The cost of land is more expensive, and there are fewer rules and regulations for structures built on bodies of water not used for recreation. Unlike most land-based solar plants, floating arrays can be unobtrusive because they are hidden from public view. They achieve higher efficiencies than PV panels on land, because water cools the panels. The panels have a special coating to prevent rust or corrosion.

In May 2008, the Far Niente Winery in Oakville, California, pioneered the world's first floatovoltaic system by installing 994 solar PV modules with a total capacity of 477 kW onto 130 pontoons and floating them on the winery's irrigation pond. The primary benefit of such a system is that it avoids the need to sacrifice valuable land area that could be used for another purpose. In the case of the Far Niente Winery, it saved three-quarters of an acre that would have been required for a land-based system. Another benefit of a floatovoltaic system is that the panels are kept at a cooler temperature than they would be on land, leading to a higher efficiency of solar energy conversion. The floating PV array also reduces the amount of water lost through evaporation and inhibits the growth of algae.

Utility-scale floating PV farms are starting to be built. The multinational electronics and ceramics manufacturer Kyocera will develop the world's largest, a 13.4 MW farm on the reservoir above Yamakura Dam in Chiba Prefecture using 50,000 solar panels. Salt-water resistant floating farms are also being considered for ocean use, with experiments in Thailand. The largest so far announced floatovoltaic project is a 350 MW power station in the Amazon region of Brazil.

Direct Current Grid

DC grids are found in electric powered transport: railways trams and trolleybuses. A few pilot plants for such applications have been built, such as the tram depots in Hannover Leinhausen, using photovoltaic contributors and Geneva (Bachet de Pesay). The 150 kW$_p$ Geneva site feeds 600V DC directly into the tram/trolleybus electricity network whereas before it provided about 15% of the electricity at its opening in 1999.

Standalone

An isolated mountain hut in Catalonia, Spain

A stand-alone or off-grid system is not connected to the electrical grid. Standalone systems vary widely in size and application from wristwatches or calculators to remote buildings or spacecraft. If the load is to be supplied independently of solar insolation, the generated power is stored and buffered with a battery. In non-portable applications where weight is not an issue, such as in buildings, lead acid batteries are most commonly used for their low cost and tolerance for abuse.

Solar parking meter in Edinburgh, Scotland

A charge controller may be incorporated in the system to avoid battery damage by excessive charging or discharging. It may also help to optimize production from the solar array using a maximum power point tracking technique (MPPT). However, in simple PV systems where the PV module voltage is matched to the battery voltage, the use of MPPT electronics is generally considered unnecessary, since the battery voltage is stable enough to provide near-maximum power collection from the PV module. In small devices (e.g. calculators, parking meters) only direct current (DC) is consumed. In larger systems (e.g. buildings, remote water pumps) AC is usually required. To convert the DC from the modules or batteries into AC, an inverter is used.

In agricultural settings, the array may be used to directly power DC pumps, without the need for an inverter. In remote settings such as mountainous areas, islands, or other places where a power grid is unavailable, solar arrays can be used as the sole source of electricity, usually by charging a storage battery. Stand-alone systems closely relate to microgeneration and distributed generation.

- Pico PV systems: The smallest, often portable photovoltaic systems are called pico solar PV systems, or pico solar. They mostly combine a rechargeable battery and charge controller, with a very small PV panel. The panel's nominal capacity is just a few watt-peak ($1-10$ W_p) and its area less than a tenth of a square meter, or one square foot, in size. A large range of different applications can be solar powered such as music players, fans, portable lamps, security lights, solar lighting kits, solar lanterns and street light, phone chargers, radios, or even small, seven-inch LCD televisions, that run on less than ten watts. As it is the case for power generation from pico hydro, pico PV systems are useful in small, rural communities that require only a small amount of electricity. Since the efficiency of many appliances have improved considerably, in particular due to the usage of LED lights and efficient rechargeable batteries, pico solar has become an affordable alternative, especially in the developing world. The metric prefix *pico-* stands for a *trillionth* to indicate the smallness of the system's electric power.

- Solar street lights: Solar street lights raised light sources which are powered by photovoltaic panels generally mounted on the lighting structure. The solar array of such off-grid PV system charges a rechargeable battery, which powers a fluorescent or LED lamp during the night. Solar street lights are stand-alone power systems, and have the advantage of savings on trenching, landscaping, and maintenance costs, as well as on the electric bills, despite their higher initial cost compared to conventional street lighting. They are designed with sufficiently large batteries to ensure operation for at least a week and even in the worst situation, they are expected to dim only slightly.

- Telecommunication and signaling: Solar PV power is ideally suited for telecommunication applications such as local telephone exchange, radio and TV broadcasting, microwave and other forms of electronic communication links. This is because, in most telecommunication application, storage batteries are already in use and the electrical system is basically DC. In hilly and mountainous terrain, radio and TV signals may not reach as they get blocked or reflected back due to undulating terrain. At these locations, low power transmitters are installed to receive and retransmit the signal for local population.

- Solar vehicles: Solar vehicle, whether ground, water, air or space vehicles may obtain some or all of the energy required for their operation from the sun. Surface vehicles generally require higher power levels than can be sustained by a practically sized solar array, so a battery assists in meeting peak power demand, and the solar array recharges it. Space vehicles have successfully used solar photovoltaic systems for years of operation, eliminating the weight of fuel or primary batteries.

- Solar pumps: One of the most cost effective solar applications is a solar powered pump, as it is far cheaper to purchase a solar panel than it is to run power lines. They often meet a need for water beyond the reach of power lines, taking the place of a windmill or windpump. One common application is the filling of livestock watering tanks, so that grazing cattle may drink. Another is the refilling of drinking water storage tanks on remote or self-sufficient homes.

- Spacecraft: Solar panels on spacecraft have been one of the first applications of photovoltaics since the launch of Vanguard 1 in 1958, the first satellite to use solar cells. Contrary to Sputnik, the first artificial satellite to orbit the planet, that ran out of batteries within 21 days due to the lack of solar-power, most modern communications satellites and space probes in the inner solar system rely on the use of solar panels to derive electricity from sunlight.

- Do it yourself community: With a growing interest in environmentally friendly green energy, an increasing number of hobbyists in the DIY-community have endeavored to build their own solar PV systems from kits or partly DIY. Usually, the DIY-community uses inexpensive or high efficiency systems (such as those with solar tracking) to generate their own power. As a result, the DIY-systems often end up cheaper than their commercial counterparts. Often, the system is also hooked up into the regular power grid, using net metering instead of a battery for backup. These systems usually generate power amount of ~2 kW or less. Through the internet, the community is now able to obtain plans to (partly) construct the system and there is a growing trend toward building them for domestic requirements.

Costs and Economy

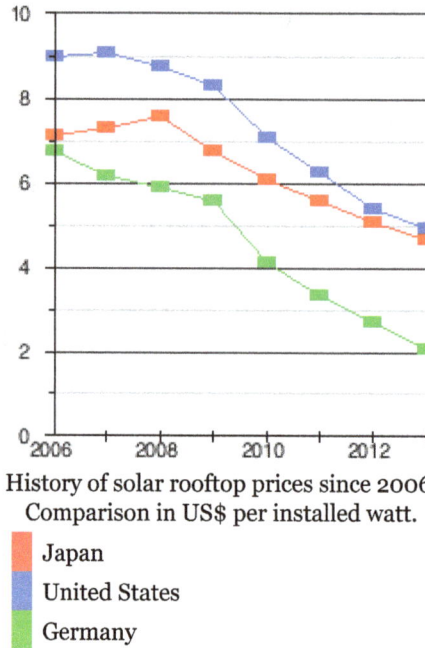

Median installed system prices for residential PV Systems in Japan, Germany and the United States ($/W)

History of solar rooftop prices since 2006. Comparison in US$ per installed watt.

Japan
United States
Germany

The cost of producing photovotaic cells have dropped due to economies of scale in production and technological advances in manufacturing. For large-scale installations, prices below $1.00 per watt were common by 2012. A price decrease of 50% had been achieved in Europe from 2006 to 2011 and there is a potential to lower the generation cost by 50% by 2020. Crystal silicon solar cells have largely been replaced by less expensive multicrystalline silicon solar cells, and thin film silicon solar cells have also been developed recently at lower costs of production. Although they are reduced in energy conversion efficiency from single crystalline "siwafers", they are also much easier to produce at comparably lower costs.

The table below shows the total cost in US cents per kWh of electricity generated by a photovoltaic system. The row headings on the left show the total cost, per peak kilowatt (kW_p), of a photovoltaic installation. Photovoltaic system costs have been declining and in Germany, for example, were reported to have fallen to USD $1389/kW_p$ by the end of 2014. The column headings across the top refer to the annual energy output in kWh expected from each installed kW_p. This varies by geographic region because the average insolation depends on the average cloudiness and the thickness of atmosphere traversed by the sunlight. It also depends on the path of the sun relative to the panel and the horizon. Panels are usually mounted at an angle based on latitude, and often they are adjusted seasonally to meet the changing solar declination. Solar tracking can also be utilized to access even more perpendicular sunlight, thereby raising the total energy output.

The calculated values in the table reflect the total cost in cents per kWh produced. They assume a 10% total capital cost (for instance 4% interest rate, 1% operating and maintenance cost, and depreciation of the capital outlay over 20 years). Normally, photovoltaic modules have a 25-year warranty.

Cost of generated kilowatt-hour by a PV-System (US¢/kWh) depending on solar radiation and installation cost during 20 years of operation									
Installation cost in $ per watt	**Insolation annually generated kilowatt-hours per installed kW-capacity (kWh/kWp·y)**								
	2400	**2200**	**2000**	**1800**	**1600**	**1400**	**1200**	**1000**	**800**
$0.20	0.8	0.9	1.0	1.1	1.3	1.4	1.7	2.0	2.5
$0.60	2.5	2.7	3.0	3.3	3.8	4.3	5.0	6.0	7.5
$1.00	4.2	4.5	5.0	5.6	6.3	7.1	8.3	10.0	12.5
$1.40	5.8	6.4	7.0	7.8	8.8	10.0	11.7	14.0	17.5
$1.80	7.5	8.2	9.0	10.0	11.3	12.9	15.0	18.0	22.5
$2.20	9.2	10.0	11.0	12.2	13.8	15.7	18.3	22.0	27.5
$2.60	10.8	11.8	13.0	14.4	16.3	18.6	21.7	26.0	32.5
$3.00	12.5	13.6	15.0	16.7	18.8	21.4	25.0	30.0	37.5
$3.40	14.2	15.5	17.0	18.9	21.3	24.3	28.3	34.0	42.5
$3.80	15.8	17.3	19.0	21.1	23.8	27.1	31.7	38.0	47.5
$4.20	17.5	19.1	21.0	23.3	26.3	30.0	35.0	42.0	52.5
$4.60	19.2	20.9	23.0	25.6	28.8	32.9	38.3	46.0	57.5
$5.00	20.8	22.7	25.0	27.8	31.3	35.7	41.7	50.0	62.5

USA	Japan	Germany	Small rooftop system cost and avg. insolation applied to data table, in 2013

Notes:

1. Cost per watt for rooftop system in 2013: Japan $4.64, United States $4.92, and Germany $2.05

2. Generated kilowatt-hour per installed watt-peak, based on average insolation for Japan (1500 kWh/m²/year), United States (5.0 to 5.5 kWh/m²/day), and Germany (1000 to 1200 kWh/m²/year).

3. A 2013 study by the Fraunhofer ISE concludes LCOE cost for a small PV system to be $0.16 (€0.12) rather than $0.22 per kilowatt-hour as shown in table (Germany).

System Cost 2013

In its 2014 edition of the "Technology Roadmap: Solar Photovoltaic Energy" report, the International Energy Agency (IEA) published prices in US$ per watt for residential, commercial and utility-scale PV systems for eight major markets in 2013.

Typical PV system prices in 2013 in selected countries (USD)								
USD/W	*Australia*	*China*	*France*	*Germany*	*Italy*	*Japan*	*United Kingdom*	*United States*
Residential	1.8	1.5	4.1	2.4	2.8	4.2	2.8	4.9
Commercial	1.7	1.4	2.7	1.8	1.9	3.6	2.4	4.5
Utility-scale	2.0	1.4	2.2	1.4	1.5	2.9	1.9	3.3

Regulation

Standardization

Increasing use of photovoltaic systems and integration of photovoltaic power into existing structures and techniques of supply and distribution increases the value of general standards and definitions for photovoltaic components and systems. The standards are compiled at the International Electrotechnical Commission (IEC) and apply to efficiency, durability and safety of cells, modules, simulation programs, plug connectors and cables, mounting systems, overall efficiency of inverters etc.

Planning and Permit

While article 690 of the National Electric Code provides general guidelines for the installation of photovoltaic systems, these guidelines may be superseded by local laws and regulations. Often a permit is required necessitating plan submissions and structural calculations before work may begin. Additionally, many locales require the work to be performed under the guidance of a licensed electrician. Check with the local City/County AHJ (Authority Having Jurisdiction) to ensure compliance with any applicable laws or regulations.

In the United States, the Authority Having Jurisdiction (AHJ) will review designs and issue permits, before construction can lawfully begin. Electrical installation practices must comply with standards set forth within the National Electrical Code (NEC) and be inspected by the AHJ to ensure compliance with building code, electrical code, and fire safety code. Jurisdictions may require that equipment has been tested, certified, listed, and labeled by at least one of the Nationally Recognized Testing Laboratories (NRTL). Despite the complicated installation process, a recent list of solar contractors shows a majority of installation companies were founded since 2000.

National Regulations

United Kingdom

In the UK, PV installations are generally considered permitted development and don't require planning permission. If the property is listed or in a designated area (National Park, Area of Outstanding Natural Beauty, Site of Special Scientific Interest or Norfolk Broads) then planning permission is required.

United States

In the US, many localities require a permit to install a photovoltaic system. A grid-tied system normally requires a licensed electrician to make the connection between the system and the grid-connected wiring of the building. Installers who meet these qualifications are located in almost every state. The State of California prohibits homeowners' associations from restricting solar devices.

Spain

Although Spain generates around 40% of its electricity via photovoltaic and other renewable energy sources, and cities such as Huelva and Seville boast nearly 3,000 hours of sunshine per year, Spain has issued a solar tax to account for the debt created by the investment done by the Spanish government. Those who do not connect to the grid can face up to a fine of 30 million euros ($40 million USD).

Grid-connected Photovoltaic Power System

A grid-connected, residential solar rooftop system near Boston, USA

A grid-connected photovoltaic power system, or grid-connected PV power system is an electricity generating solar PV power system that is connected to the utility grid. A grid-connected PV system consists of solar panels, one or several inverters, a power conditioning unit and grid connection equipment. They range from small residential and commercial rooftop systems to large utility-scale solar power stations. Unlike stand-alone power systems, a grid-connected system rarely includes an integrated battery solution, as they are still very expensive. When conditions are right, the grid-connected PV system supplies the excess power, beyond consumption by the connected load, to the utility grid.

Operation

Residential, grid-connected rooftop systems which have a capacity more than 10 kilowatts can meet the load of most consumers. They can feed excess power to the grid where it is consumed by other users. The feedback is done through a meter to monitor power transferred. Photovoltaic wattage may be less than average consumption, in which case the consumer will continue to purchase grid energy, but a lesser amount than previously. If photovoltaic wattage substantially exceeds average consumption, the energy produced by the panels will be much in excess of the demand. In this case, the excess power can yield revenue by selling it to the grid. Depending on their agreement with their local grid energy company, the consumer only needs to pay the cost of electricity consumed less the value of electricity generated. This will be a negative number if more electricity is generated than consumed. Additionally, in some cases, cash incentives are paid from the grid operator to the consumer.

Connection of the photovoltaic power system can be done only through an interconnection agreement between the consumer and the utility company. The agreement details the various safety standards to be followed during the connection.

Features

Solar energy gathered by photovoltaic solar panels, intended for delivery to a power grid, must be conditioned, or processed for use, by a grid-connected inverter. Fundamentally, an inverter changes the DC input voltage from the PV to AC voltage for the grid. This inverter sits between the solar array and the grid, draws energy from each, and may be a large stand-alone unit or may be a collection of small inverters, each physically attached to individual solar panels. The inverter must monitor grid voltage, waveform, and frequency. One reason for monitoring is if the grid is dead or strays too far out of its nominal specifications, the inverter must not pass along any solar energy. An inverter connected to a malfunctioning power line will automatically disconnect in accordance with safety rules, for example UL1741, which vary by jurisdiction. Another reason for the inverter monitoring the grid is because for normal operation the inverter must synchronize with the grid waveform, and produce a voltage slightly higher than the grid itself, in order for energy to smoothly flow outward from the solar array.

Anti-islanding

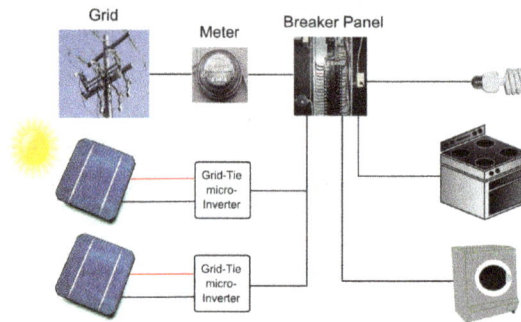

Diagram of a residential grid-connected PV system

Islanding is the condition in which a distributed generator continues to power a location even though power from the electric utility grid is no longer present. Islanding can be dangerous to utility workers, who may not realize that a circuit is still powered, even though there is no power from the electrical grid. For that reason, distributed generators must detect islanding and immediately stop producing power; this is referred to as anti-islanding.

In the case of a utility blackout in a grid-connected PV system, the solar panels will continue to deliver power as long as the sun is shining. In this case, the supply line becomes an "island" with power surrounded by a "sea" of unpowered lines. For this reason, solar inverters that are designed to supply power to the grid are generally required to have automatic anti-islanding circuitry in them. In intentional islanding, the generator disconnects from the grid, and forces the distributed generator to power the local circuit. This is often used as a power backup system for buildings that normally sell their power to the grid.

There are two types of anti-islanding control techniques:

- *Passive:* The voltage and/or the frequency change during the grid failure is measured and a positive feedback loop is employed to push the voltage and/or the frequency further away from its nominal value. Frequency or voltage may not change if the load matches very well with the inverter output or the load has a very high quality factor (reactive to real power ratio). So there exists some *Non Detection Zone* (NDZ).

- *Active:* This method employs injecting some error in frequency or voltage. When grid fails, the error accumulates and pushes the voltage and/or frequency beyond the acceptable range.

Advantages

- Systems such as Net Metering and Feed-in Tariff which are offered by some system operators, can offset a customers electricity usage costs. In some locations though, grid technologies cannot cope with distributed generation feeding into the grid, so the export of surplus electricity is not possible and that surplus is earthed.

- Grid-connected PV systems are comparatively easier to install as they do not require a battery system.

- Grid interconnection of photovoltaic (PV) power generation systems has the advantage of effective utilization of generated power because there are no storage losses involved.

- A photovoltaic power system is carbon negative over its lifespan, as any energy produced over and above that to build the panel initially offsets the need for burning fossil fuels. Even though the sun doesn't always shine, any installation gives a reasonably predictable average reduction in carbon consumption.

Disadvantages

- Grid-connected PV can cause issues with voltage regulation. The traditional grid operates under the assumption of one-way, or radial, flow. But electricity injected into the grid increases voltage, and can drive levels outside the acceptable bandwidth of ±5%.

- Grid-connected PV can compromise power quality. PV's intermittent nature means rapid changes in voltage. This not only wears out voltage regulators due to frequent adjusting, but also can result in voltage flicker.

- Connecting to the grid poses many protection-related challenges. In addition to islanding, as mentioned above, too high levels of grid-connected PV result in problems like relay desensitization, nuisance tripping, interference with automatic reclosers, and ferroresonance.

Concentrator Photovoltaics

This Amonix system in Las Vegas, USA consists of thousands of small Fresnel lenses, each focusing sunlight to ~500X higher intensity onto a tiny, high-efficiency multi-junction solar cell.
A Tesla Roadster is parked beneath for scale.

Concentrator photovoltaics (CPV) modules on dual axis solar trackers in Golmud, China.

Concentrator photovoltaics (CPV) (also known as Concentration Photovoltaics) is a photovoltaic technology that generates electricity from sunlight. Contrary to conventional photovoltaic systems, it uses lenses and curved mirrors to focus sunlight onto small, but highly efficient, multi-junction (MJ) solar cells. In addition, CPV systems often use solar trackers and sometimes a cooling system to further increase their efficiency. Ongoing research and development is rapidly improving their competitiveness in the utility-scale segment and in areas of high insolation. This sort of solar technology can be thus used in smaller areas.

Systems using high-concentration photovoltaics (HCPV) especially have the potential to become competitive in the near future. They possess the highest efficiency of all existing PV technologies, and a smaller photovoltaic array also reduces the balance of system costs. Currently, CPV is not used in the PV rooftop segment and is far less common than conventional PV systems. For regions with a high annual direct normal irradiance of 2000 kilowatt-hour (kWh) per square meter or more, the levelized cost of electricity is in the range of $0.08–$0.15 per kWh and installation cost for a 10-megawatt CPV power plant was identified to lie between €1.40–€2.20 (~$1.50-$2.30) per watt-peak (W_p).

In 2016, cumulative CPV installations reached 350 megawatts (MW), less than 0.2% of the global installed capacity of 230,000 MW. Commercial HCPV systems reached instantaneous ("spot") efficiencies of up to 42% under standard test conditions (with concentration levels above 400) and the International Energy Agency sees potential to increase the efficiency of this technology to 50% by the mid-2020s. As of December 2014, the best lab cell efficiency for concentrator MJ-cells reached 46% (four or more junctions). Under outdoor, operating conditions, CPV module efficiencies have exceeded 33% ("one third of a sun"). System-level AC efficiencies are in the range of 25-28%. CPV installations are located in China, the United States, South Africa, Italy and Spain.

HCPV directly competes with concentrated solar power (CSP) as both technologies are suited best for areas with high direct normal irradiance, which are also known as the Sun Belt region in the United States and the Golden Banana in Southern Europe. CPV and CSP are often confused with one another, despite being intrinsically different technologies from the start: CPV uses the photovoltaic effect to directly generate electricity from sunlight, while CSP – often called *concentrated solar thermal* – uses the heat from the sun's radiation in order to make steam to drive a turbine, that then produces electricity using a generator. Currently, CSP is more common than CPV.

History

Research into concentrator photovoltaics has taken place since the mid 1970s, initially spurred on by the energy shock from a mideast oil embargo. Sandia National Laboratories in Albuquerque, New Mexico was the site for most of the early work, with the first modern-like photovoltaic concentrating system produced there late in the decade. Their first system was a linear-trough concentrator system that used a point focus acrylic Fresnel lens focusing on water-cooled silicon cells and two axis tracking. Cell cooling with a passive heat sink was demonstrated in 1979 by Ramón Areces. The 350 kW SOLERAS project in Saudi Arabia - the largest until many years later - was constructed by Sandia/Martin Marietta in 1981.

Research and development continued through the 1980's and 1990's without significant industry interest. Improvements in cell efficiency were soon recognized as essential to making the technology economical. However the improvements to Si-based cell technologies used by both concentrators and flat PV failed to favor the system-level economics of CPV. The introduction of III-V Multi-junction solar cells starting in the early 2000's has since provided a clear differentiator. MJ cell efficiencies have improved from 34% (3-junctions) to 46% (4-junctions) at research-scale production levels. A substantial number of multi-MW CPV projects have also been commissioned worldwide since 2010.

Challenges

Modern CPV systems operate most efficiently in highly concentrated sunlight (i.e. concentration levels equivalent to hundreds of suns), as long as the solar cell is kept cool through the use of heat sinks. Diffuse light, which occurs in cloudy and overcast conditions, cannot be highly concentrated using conventional optical components only (i.e. macroscopic lenses and mirrors). Filtered light, which occurs in hazy or polluted conditions, has spectral variations which produce mismatches between the electrical currents generated within the series-connected junctions of spectrally "tuned" multi-junction (MJ) photovoltaic cells. These CPV features lead to rapid decreases in power output when atmospheric conditions are less than ideal.

To produce equal or greater energy per rated watt than conventional PV systems, CPV systems must be located in areas that receive plentiful direct sunlight. This is typically specified as average DNI greater than 5.5-6 kWh/m²/day or 2000kWh/m²/yr. Otherwise, evaluations of annualized DNI vs. GNI/GHI irradiance data have concluded that conventional PV should still perform better over time than presently available CPV technology in most regions of the world.

In late 2015, ARPA-E announced a first round of R&D funding for the MOSAIC Program (Microscale Optimized Solar-cell Arrays with Integrated Concentration) to further combat the location and expense challenges of existing CPV technology. As stated in the program description: "MOSAIC projects are grouped into three categories: complete systems that cost effectively integrate micro-CPV for regions such as sunny areas of the U.S. southwest that have high Direct Normal Incident (DNI) solar radiation; complete systems that apply to regions, such as areas of the U.S. Northeast and Midwest, that have low DNI solar radiation or high diffuse solar radiation; and concepts that seek partial solutions to technology challenges."

In Europe the CPVMATCH Program (Concentrating PhotoVoltaic Modules using Advanced Technologies and Cells for Highest efficiencies) aims "to bring practical performance of HCPV modules closer to theoretical limits". Efficiency goals achievable by 2019 are identified as 48% for cells and 40% for modules at >800x concentration.

CPV Strengths	CPV Weaknesses
High efficiencies under direct normal irradiance	HCPV cannot utilize diffuse radiation. LCPV can only utilize a fraction of diffuse radiation.
Low cost per watt of manufacturing capital	Power output of MJ solar cells is more sensitive to shifts in radiation spectra caused by changing atmospheric conditions.
Low temperature coefficients	Tracking with sufficient accuracy and reliability is required.
No cooling water required for passively cooled systems	May require frequent cleaning to mitigate soiling losses, depending on the site
Additional use of waste heat possible for systems with active cooling possible (e.g.large mirror systems)	Limited market – can only be used in regions with high DNI, cannot be easily installed on rooftops
Modular – kW to GW scale	Strong cost decrease of competing technologies for electricity production
Increased and stable energy production throughout the day due to (two-axis) tracking	Bankability and perception issues
Low energy payback time	New generation technologies, without a history of production (thus increased risk)
Potential double use of land e.g. for agriculture, low environmental impact	Optical losses
High potential for cost reduction	Lack of technology standardization
Opportunities for local manufacturing	–
Smaller cell sizes could prevent large fluctuations in module price due to variations in semiconductor prices	–
Greater potential for efficiency increase in the future compared to single-junction flat plate systems could lead to greater improvements in land area use, BOS costs, and BOP costs	–

Optical Design

The design of macroscopic sunlight concentrators for CPV introduces a very specific optical design problem, with features that makes it different from any other optical design. It has to be efficient, suitable for mass production, capable of high concentration, insensitive to manufacturing and mounting inaccuracies, and capable of providing uniform illumination of the cell. All these reasons make nonimaging optics the most suitable for CPV.

For very low concentrations, the wide acceptance angles of nonimaging optics avoid the need for active solar tracking. For medium and high concentrations, a wide acceptance angle can be seen as a measure of how tolerant the optic is to imperfections in the whole system. It is vital to start with a wide acceptance angle since it must be able to accommodate tracking errors, movements of the system due to wind, imperfectly manufactured optics, imperfectly assembled components, finite stiffness of the supporting structure or its deformation due to aging, among other factors. All of these reduce the initial acceptance angle and, after they are all factored in, the system must still be able to capture the finite angular aperture of sunlight.

Efficiency

All CPV systems have a concentrating optic and a solar cell. Generally, active solar tracking is necessary. Low-concentration systems often have a simple booster reflector, which can increase solar electric output by over 30% from that of non-concentrator PV systems. Experimental results from such LCPV systems in Canada resulted in energy gains over 40% for prismatic glass and 45% for traditional crystalline silicon PV modules.

Semiconductor properties allow solar cells to operate more efficiently in concentrated light, as long as the cell Junction temperature is kept cool by suitable heat sinks. Efficiency of multi-junction photovoltaic cells developed in research is upward of 44% today, with the potential to approach 50% in the coming years. The theoretical limiting efficiency under concentration approaches 65% for 5 junctions, which is a likely practical maximum.

Types

CPV systems are categorized according to the amount of their solar concentration, measured in "suns" (the square of the magnification).

Low Concentration PV (LCPV)

Low concentration PV are systems with a solar concentration of 2–100 suns. For economic reasons, conventional or modified silicon solar cells are typically used, and, at these concentrations, the heat flux is low enough that the cells do not need to be actively cooled. There is now modeling and experimental evidence that standard solar modules do not need any modification, tracking or cooling if the concentration level is low and yet still have increased output of 35% or more. The laws of optics dictate that a solar collector with a low concentration ratio can have a high acceptance angle and thus in some instances does not require active solar tracking.

An example of a Low Concentration PV Cell's surface, showing the glass lensing.

Medium Concentration PV

From concentrations of 100 to 300 suns, the CPV systems require two-axes solar tracking and cooling (whether passive or active), which makes them more complex.

A 10×10 mm HCPV solar cell

High Concentration Photovoltaics (HCPV)

High concentration photovoltaics (HCPV) systems employ concentrating optics consisting of dish reflectors or fresnel lenses that concentrate sunlight to intensities of 1,000 suns or more. The solar cells require high-capacity heat sinks to prevent thermal destruction and to manage temperature related electrical performance and life expectancy losses. To further exacerbate the concentrated cooling design, the heat sink must be passive, otherwise the power required for active cooling will reduce the overall conversion efficiency and economy. Multi-junction solar cells are currently favored over single junction cells, as they are more efficient and have a lower temperature coefficient (less loss in efficiency with an increase in temperature). The efficiency of both cell types rises with increased concentration; multi-junction efficiency rises faster. Multi-junction solar cells, originally designed for non-concentrating PV on space-based satellites, have been re-designed due to the high-current density encountered with CPV (typically 8 A/cm^2 at 500 suns). Though the cost of multi-junction solar cells is roughly 100 times that of conventional silicon cells of the same area, the small cell area employed makes the relative costs of cells in each system comparable and the system economics favor the multi-junction cells. Multi-junction cell efficiency has now reached 44% in production cells.

The 44% value given above is for a specific set of conditions known as "standard test conditions". These include a specific spectrum, an incident optical power of 850 W/m^2, and a cell temperature of 25 °C. In a concentrating system, the cell will typically operate under conditions of variable spectrum, lower optical power, and higher temperature. The optics needed to concentrate the light have limited efficiency themselves, in the range of 75–90%. Taking these factors into account, a solar module incorporating a 44% multi-junction cell might deliver a DC efficiency around 36%. Under similar conditions, a crystalline silicon module would deliver an efficiency of less than 18%.

When high concentration is needed (500–1000 times), as occurs in the case of high efficiency multi-junction solar cells, it is likely that it will be crucial for commercial success at the system level to achieve such concentration with a sufficient acceptance angle. This allows tolerance in mass production of all components, relaxes the module assembling and system installation, and decreasing the cost of structural elements. Since the main goal of CPV is to make solar energy inexpensive, there can be used only a few surfaces. Decreasing the number of elements and achieving high acceptance angle, can be relaxed optical and mechanical requirements, such as accuracy of the optical surfaces profiles, the module assembling, the installation, the supporting structure, etc. To this end, improvements in sunshape modelling at the system design stage may lead to higher system efficiencies.

Installations

Concentrator photovoltaics technology has established its presence in the solar industry over the last several years. The first CPV power plant that exceeded 1 MW-level was commissioned in Spain in 2006. By the end of 2015, the number of CPV power plants around the world accounted for a total installed capacity of 350 MW. Field data collected over the past six years is also starting to benchmark the prospects for long-term system reliability.

The emerging CPV segment has comprised ~0.1% of the fast-growing utility market for PV installations over the past decade. Unfortunately, by the end of 2015, the near term outlook for CPV industry growth has faded with closure of all of the largest CPV manufacturing facilities: including those of Suncore, Soitec, Amonix, and Solfocus. Nevertheless, the growth outlook for the overall PV industry continues to appear strong.

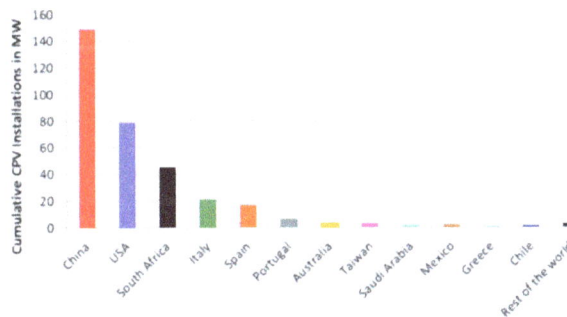

Cumulative CPV Installations in MW by country by November 2014

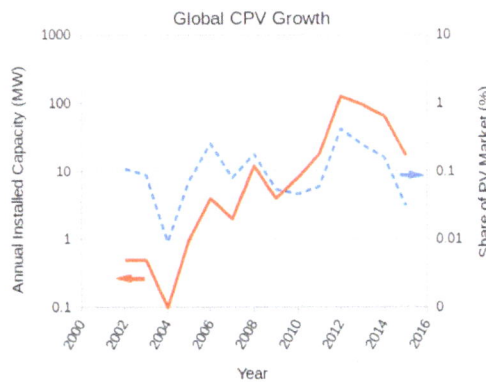

Yearly Installed CPV Capacity in MW from 2002 to 2015

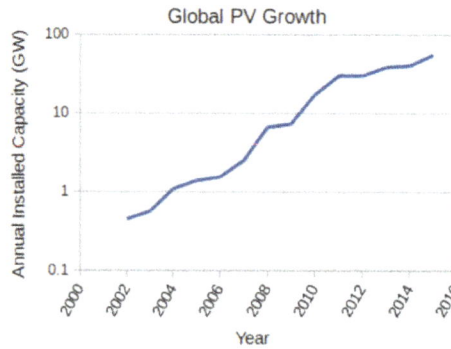

Yearly Installed PV Capacity in GW from 2002 to 2015

List of large CPV Systems

The largest CPV power plant currently in operation is of 80 MW_p capacity located in Golmud, China, hosted by Suncore Photovoltaics.

Power station	Capacity (MW$_p$)	Location	Vendor/Builder
Golmud 2	79.83	in Golmud/Qinghai province/China	Suncore
Golmud 1	57.96	in Golmud/Qinghai province/China	Suncore
Touwsrivier	44.19	in Touwsrivier/Western Cape/South Africa	Soitec
Alamosa Solar Project	35.28	in Alamosa, Colorado/San Luis Valley/United States	Amonix
Source: The CPV Consortium			

Concentrated Photovoltaics and Thermal

Concentrator photovoltaics and thermal (CPVT), also sometimes called combined heat and power solar (CHAPS) or hybrid thermal CPV, is a cogeneration or micro cogeneration technology used in the field of concentrator photovoltaics that produces usable heat and electricity within the same system. CPVT at high concentrations of over 100 suns (HCPVT) utilizes similar components as HCPV, including dual-axis tracking and multi-junction photovoltaic cells. A fluid actively cools the integrated thermal–photovoltaic receiver, and simultaneously transports the collected heat.

Typically, one or more receivers and a heat exchanger operate within a closed thermal loop. To maintain efficient overall operation and avoid damage from thermal runaway, the demand for heat from the secondary side of the exchanger must be consistently high. Under such optimal operating conditions, collection efficiencies exceeding 70% (up to ~35% electric, ~40% thermal for HCPVT) are anticipated. Net operating efficiencies may be substantially lower depending on how well a system is engineered to match the demands of the particular thermal application.

The maximum temperature of CPVT systems is typically too low alone to power a boiler for additional steam-based cogeneration of electricity. Such systems may be economical to power lower temperature applications having a constant high heat demand. The heat may be employed in district heating, water heating and air conditioning, desalination or process heat. For applications having lower or intermittent heat demand, a system may be augmented with a switchable heat dump to the external environment in order to maintain reliable electrical output and safeguard cell life, despite the resulting reduction in net operating efficiency.

HCPVT active cooling enables the use of much higher power thermal–photovoltaic receiver units, generating typically 1–100 kilowatts electric, as compared to HCPV systems that mostly rely upon passive cooling of single ~20W cells. Such high-power receivers utilize dense arrays of cells mounted on a high-efficiency heat sink. Minimizing the number of individual receiver units is a simplification that should ultimately yield improvement in the overall balance of system costs, manufacturability, maintainability/upgradeability, and reliability.

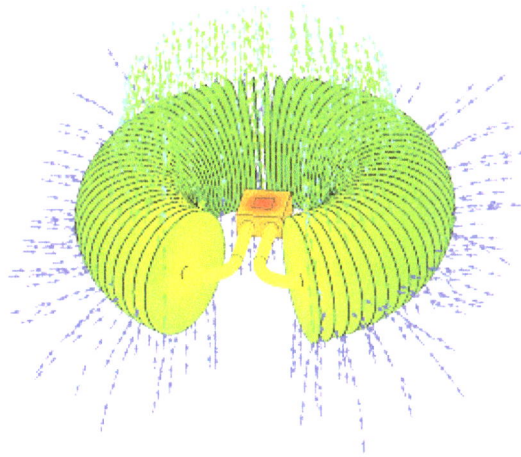

This 240 x 80 mm 1,000 suns CPV heat sink design, was created using high resolution CFD analysis, and shows temperature contoured heat sink surface and flow trajectories as predicted.

Reliability Requirements

The maximum operating temperatures ($T_{max\,cell}$) of CPVT systems are limited to less than approximately 100–125 °C on account of the intrinsic reliability limitation of their multi-junction PV cells. This contrasts to CSP and other CHP systems which may be designed to function at temperatures in excess of several hundred degrees. More specifically, the multi-junction photovoltaic cells are fabricated from a layering of thin-film III-V semiconductor materials having intrinsic lifetimes during CPV operation that rapidly decrease with an Arrhenius-type temperature dependence. The system receiver must therefore provide for highly efficient and uniform cell cooling, where an ideal receiver would provide $T_{max\,coolant} \sim T_{max\,cell}$. In addition to material and design limitations in receiver heat-transfer performance, numerous extrinsic factors, such as the frequent system thermal cycling, further reduce the practical $T_{max\,coolant}$ compatible with long system life to below about 80 °C.

The higher capital costs, lesser standardization, and added engineering & operational complexities (in comparison to zero and low-concentration PV technologies) make demonstrations of system reliability and long-life performance critical challenges for the first generation of CPV and CPVT technologies. Performance certification testing standards (e.g. IEC 62108, UL 8703, IEC 62789, IEC 62670) include stress conditions that may be useful to uncover some predominantly infant and early life (<1–2 year) failure modes at the system, module, and sub-component levels. However, such standardized tests – as typically performed on only a small sampling of units – are generally incapable to evaluate comprehensive long-term (10 to 25 or more years) lifetimes for each unique CPVT system design and application under its broader range of actual operating conditions. Long-life performance of these complex systems is therefore assessed in the field, and is improved through aggressive product development cycles which are guided by the results of accelerated component/system aging, enhanced performance monitoring diagnostics, and failure analysis. Significant growth in the deployment of CPV and CPVT can be anticipated once the long-term performance and reliability concerns are better addressed to build confidence in system bankability.

Demonstration Projects

The economics of a mature CPVT industry is anticipated to be competitive, despite the large recent cost reductions and gradual efficiency improvements for conventional silicon PV (which can be installed alongside conventional CSP to provide for similar electrical+thermal generation capabilities). CPVT may currently be economical for niche markets having all of the following application characteristics:

- high solar direct normal incidence (DNI)

- tight space constraints for placement of a solar collector array

- high and constant demand for low-temperature (<80 °C) heat

- high cost of grid electricity

- access to backup sources of power or cost-efficient storage (electrical and thermal)

Utilization of a power purchase agreement (PPA), government assistance programs, and innovative financing schemes are also helping potential manufacturers and users to mitigate the risks of early CPVT technology adoption.

CPVT equipment offerings ranging from low (LCPVT) to high (HCPVT) concentration are now being deployed by several startup ventures. As such, longer-term viability of the technical and/or business approach being pursued by any individual system provider is typically speculative. Notably, the minimum viable products of startups can vary widely in their attention to reliability engineering. Nevertheless, the following incomplete compilation is offered to assist with the identification of some early industry trends.

LCPVT systems at ~14x concentration using reflective trough concentrators, and receiver pipes clad with silicon cells having dense interconnects, have been assembled by Cogenra with a claimed 75% efficiency (~15-20% electric, 60% thermal). Several such systems are in operation for more than 5 years as of 2015, and similar systems are being produced by Absolicon and Idhelio at 10x and 50x concentration, respectively.

HCPVT offerings at over 700x concentration have more recently emerged, and may be classified into three power tiers. Third tier systems are distributed generators consisting of large arrays of ~20W single-cell receiver/collector units, similar to those previously pioneered by Amonix and SolFocus for HCPV. Second tier systems utilize localized dense-arrays of cells that produce 1-100 kW of electrical power output per receiver/generator unit. First tier systems exceed 100 kW of electrical output and are most aggressive in targeting the utility market.

Several HCPVT system providers are listed in the following table. Nearly all are early demonstration systems which have been in service for under 5 years as of 2015. Collected thermal power is typically 1.5x-2x the rated electrical power.

Provider	Country	Concentrator Type	Unit Size in kW$_e$	
			Generator	Receiver
		- Tier 1 -		
Raygen	Australia	Large Heliostat Array	200	200
Sunfish	United Kingdom	Large Heliostat Array	na	na
		- Tier 2 -		
Renovalia	Spain	Large Dish	3	3
Zenith Solar/Suncore	Israel/China/USA	Large Dish	4.5	2.25
Sun Oyster	Germany	Large Trough + Lens	4.7	2.35
Rehnu	United States	Large Dish	6.4	0.8
Forbes Solar	India/Germany	Large Dish	7.5	3.75
Airlight Energy/dsolar	Switzerland	Large Dish	12	12
Southwest Solar	United States	Large Dish	20	20
Silex Solar Systems	Australia	Large Dish	35	35
		- Tier 3 -		
Brightleaf	United States	Small Dish Array	4	0.02
Silex Power	Malta	Small Dish Array	16	0.04
Solergy	Italy/USA	Small Lens Array	20	0.02

Photovoltaic Thermal Hybrid Solar Collector

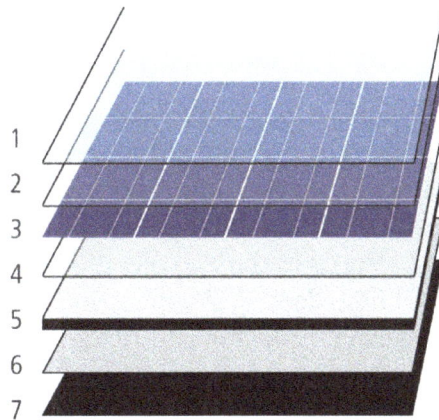

Schematic of a hybrid (PVT) solar collector: 1 - Anti-reflective glass 2 - EVA-encapsulant 3 - Solar PV cells 4 - EVA-encapsulant 5 - Backsheet (PVF) 6 - Heat exchanger (copper) 7 - Insulation (polyurethane)

Photovoltaic thermal hybrid solar collectors, sometimes known as hybrid PV/T systems or PVT, are systems that convert solar radiation into thermal and electrical energy. These systems combine a solar cell, which converts sunlight into electricity, with a solar thermal collector, which captures the remaining energy and removes waste heat from the PV module. The capture of both electricity and heat allow these devices to have higher exergy and thus be more overall energy efficient than solar photovoltaic (PV) or solar thermal alone. A significant amount of research has gone into developing PVT technology since the 1970s.

Photovoltaic cells suffer from a drop in efficiency with the rise in temperature due to increased resistance. Such systems can be engineered to carry heat away from the PV cells thereby cooling the cells and thus improving their efficiency by lowering resistance. Although this is an effective method, it causes the thermal component to under-perform compared to a solar thermal collector. Recent research showed that photovoltaic materials with low temperature coefficients such as amorphous silicon (a-Si:H) PV allow the PVT to be operated at high temperatures, creating a more symbiotic PVT system. This advantage can be tuned by controlling the dispatch strategy of thermal annealing cycles in any region of the world.

System Types

A number of PV/T collectors in different categories are commercially available and can be divided into the following categories:

- PV/T liquid collector

- PV/T air collector

- PV/Ta Liquid and air collector

- PV/T concentrator (CPVT)

PV/T Liquid Collector

The basic water-cooled design uses conductive-metal piping or plates attached to the back of a PV module. The fluid flow arrangement through the cooling element will determine which systems the panels are most suited to.

In a standard fluid-based system, a working fluid, typically water, glycol or mineral oil is then piped through these pipes or plate chillers. The heat from the PV cells is conducted through the metal and absorbed by the working fluid (presuming that the working fluid is cooler than the operating temperature of the cells). In closed-loop systems this heat is either exhausted (to cool it), or transferred at a heat exchanger, where it flows to its application. In open-loop systems, this heat is used, or exhausted before the fluid returns to the PV cells. It is also possible to disperse nanoparticles in the liquid to create a liquid filter for PV/T applications. The basic advantage of this type of split configuration is that the thermal collector and the photovoltaic collector can operate at different temperatures.

PV/T Air Collector

The basic air-cooled design uses a hollow, conductive metal housing to mount the photo-voltaic (PV) panels. Heat is radiated from the panels into the enclosed space, where the air is either circulated into a building HVAC system to recapture heat energy, or rises and is vented from the top of the structure.

While energy transfer to air is not as efficient as a liquid collector, the infrastructure required has lower cost and complexity; basically a shallow metal box. Placement of PV panels can be vertical or angled.

PV/T Concentrator (CPVT)

A concentrator system has the advantage to reduce the amount of photovoltaic (PV) cells needed, such that somewhat more expensive and efficient multi-junction photovoltaic cells can be used that will maximize the ratio of produced high-value electrical power versus lower-value thermal power. A major limitation of high-concentrator (i.e. HCPV and HCPVT) systems is that they maintain their advantage over conventional c-Si/mc-Si collectors only in regions that remain consistently free of atmospheric aerosol contaminants (e.g. light clouds, smog, etc.). Concentrator system performance is especially degraded because 1) radiation is reflected and scattered outside of the small (often less than 1°-2°) acceptance angle of the collection optics, and 2) absorption of specific components of the solar spectrum causes one or more series junctions within the MJ cells to underperfom.

Concentrator systems also require reliable control systems to accurately track the sun and to protect the PV cells from damaging over-temperature conditions. Under ideal conditions, about 75% of the suns power directly incident upon such systems can be gathered as electricity and heat.

Solar Cell

A solar cell, or photovoltaic cell (previously termed "solar battery"), is an electrical device that converts the energy of light directly into electricity by the photovoltaic effect, which is a physical and chemical phenomenon. It is a form of photoelectric cell, defined as a device whose electrical characteristics, such as current, voltage, or resistance, vary when exposed to light. Solar cells are the building blocks of photovoltaic modules, otherwise known as solar panels.

Solar cells are described as being photovoltaic, irrespective of whether the source is sunlight or an artificial light. They are used as a photodetector (for example infrared detectors), detecting light or other electromagnetic radiation near the visible range, or measuring light intensity.

The operation of a photovoltaic (PV) cell requires three basic attributes:

- The absorption of light, generating either electron-hole pairs or excitons.

- The separation of charge carriers of opposite types.

- The separate extraction of those carriers to an external circuit.

In contrast, a solar thermal collector supplies heat by absorbing sunlight, for the purpose of either direct heating or indirect electrical power generation from heat. A "photoelectrolytic cell" (photoelectrochemical cell), on the other hand, refers either to a type of photovoltaic cell (like that developed by Edmond Becquerel and modern dye-sensitized solar cells), or to a device that splits water directly into hydrogen and oxygen using only solar illumination.

Applications

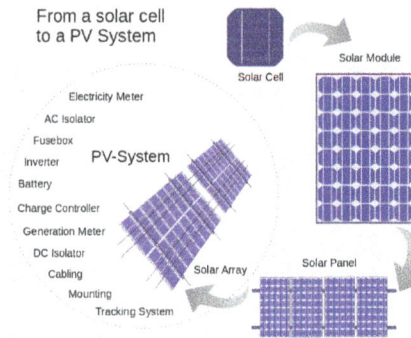

From a solar cell to a PV system. Diagram of the possible components of a photovoltaic system.

Assemblies of solar cells are used to make solar modules that generate electrical power from sunlight, as distinguished from a "solar thermal module" or "solar hot water panel". A solar array generates solar power using solar energy.

Cells, Modules, Panels and Systems

Multiple solar cells in an integrated group, all oriented in one plane, constitute a solar photovoltaic panel or solar photovoltaic module. Photovoltaic modules often have a sheet of glass on the sun-facing side, allowing light to pass while protecting the semiconductor wafers. Solar cells are usually connected in series and parallel circuits or series in modules, creating an additive voltage. Connecting cells in parallel yields a higher current; however, problems such as shadow effects can shut down the weaker (less illuminated) parallel string (a number of series connected cells) causing substantial power loss and possible damage because of the reverse bias applied to the shadowed cells by their illuminated partners. Strings of series cells are usually handled independently and not connected in parallel, though as of 2014, individual power boxes are often supplied for each module, and are connected in parallel. Although modules can be interconnected to create an array with the desired peak DC voltage and loading current capacity, using independent MPPTs (maximum power point trackers) is preferable. Otherwise, shunt diodes can reduce shadowing power loss in arrays with series/parallel connected cells.

Typical PV system prices in 2013 in selected countries (USD)								
USD/W	Australia	China	France	Germany	Italy	Japan	United Kingdom	United States
Residential	1.8	1.5	4.1	2.4	2.8	4.2	2.8	4.9
Commercial	1.7	1.4	2.7	1.8	1.9	3.6	2.4	4.5
Utility-scale	2.0	1.4	2.2	1.4	1.5	2.9	1.9	3.3

History

The photovoltaic effect was experimentally demonstrated first by French physicist Edmond Becquerel. In 1839, at age 19, he built the world's first photovoltaic cell in his father's laboratory. Willoughby Smith first described the "Effect of Light on Selenium during the passage of an Electric Current" in a 20 February 1873 issue of Nature. In 1883 Charles Fritts built the first solid state photovoltaic cell by coating the semiconductor selenium with a thin layer of gold to form the junctions; the device was only around 1% efficient.

In 1888 Russian physicist Aleksandr Stoletov built the first cell based on the outer photoelectric effect discovered by Heinrich Hertz in 1887.

In 1905 Albert Einstein proposed a new quantum theory of light and explained the photoelectric effect in a landmark paper, for which he received the Nobel Prize in Physics in 1921.

Vadim Lashkaryov discovered p-n-junctions in Cu_2O and silver sulphide protocells in 1941.

Russell Ohl patented the modern junction semiconductor solar cell in 1946 while working on the series of advances that would lead to the transistor.

The first practical photovoltaic cell was publicly demonstrated on 25 April 1954 at Bell Laboratories. The inventors were Calvin Souther Fuller and Gerald Pearson.

Solar cells gained prominence with their incorporation onto the 1958 Vanguard I satellite.

Space Applications

Solar cells were first used in a prominent application when they were proposed and flown on the Vanguard satellite in 1958, as an alternative power source to the primary battery power source. By adding cells to the outside of the body, the mission time could be extended with no major changes to the spacecraft or its power systems. In 1959 the United States launched Explorer 6, featuring large wing-shaped solar arrays, which became a common feature in satellites. These arrays consisted of 9600 Hoffman solar cells.

By the 1960s, solar cells were (and still are) the main power source for most Earth orbiting satellites and a number of probes into the solar system, since they offered the best power-to-weight ratio. However, this success was possible because in the space application, power system costs could be high, because space users had few other power options, and were willing to pay for the best possible cells. The space power market drove the development of higher efficiencies in solar cells up until the National Science Foundation "Research Applied to National Needs" program began to push development of solar cells for terrestrial applications.

In the early 1990s the technology used for space solar cells diverged from the silicon technology used for terrestrial panels, with the spacecraft application shifting to gallium arsenide-based III-V semiconductor materials, which then evolved into the modern III-V multijunction photovoltaic cell used on spacecraft.

Price Reductions

Improvements were gradual over the 1960s. This was also the reason that costs remained high, because space users were willing to pay for the best possible cells, leaving no reason to invest in lower-cost, less-efficient solutions. The price was determined largely by the semiconductor industry; their move to integrated circuits in the 1960s led to the availability of larger boules at lower relative prices. As their price fell, the price of the resulting cells did as well. These effects lowered 1971 cell costs to some $100 per watt.

In late 1969 Elliot Berman joined Exxon's task force which was looking for projects 30 years in the future and in April 1973 he founded Solar Power Corporation, a wholly owned subsidiary of Exxon at that time. The group had concluded that electrical power would be much more expensive by 2000, and felt that this increase in price would make alternative energy sources more attractive. He conducted a market study and concluded that a price per watt of about $20/watt would create significant demand. The team eliminated the steps of polishing the wafers and coating them with an anti-reflective layer, relying on the rough-sawn wafer surface. The team also replaced the expensive materials and hand wiring used in space applications with a printed circuit board on the back, acrylic plastic on the front, and silicone glue between the two, "potting" the cells. Solar cells could be made using cast-off material from the electronics market. By 1973 they announced a product, and SPC convinced Tideland Signal to use its panels to power navigational buoys, initially for the U.S. Coast Guard.

Research and Industrial Production

Research into solar power for terrestrial applications became prominent with the U.S. National Science Foundation's Advanced Solar Energy Research and Development Division within the "Research Applied to National Needs" program, which ran from 1969 to 1977, and funded research on developing solar power for ground electrical power systems. A 1973 conference, the "Cherry Hill Conference", set forth the technology goals required to achieve this goal and outlined an ambitious project for achieving them, kicking off an applied research program that would be ongoing for several decades. The program was eventually taken over by the Energy Research and Development Administration (ERDA), which was later merged into the U.S. Department of Energy.

Following the 1973 oil crisis, oil companies used their higher profits to start (or buy) solar firms, and were for decades the largest producers. Exxon, ARCO, Shell, Amoco (later purchased by BP) and Mobil all had major solar divisions during the 1970s and 1980s. Technology companies also participated, including General Electric, Motorola, IBM, Tyco and RCA.

Declining Costs and Exponential Growth

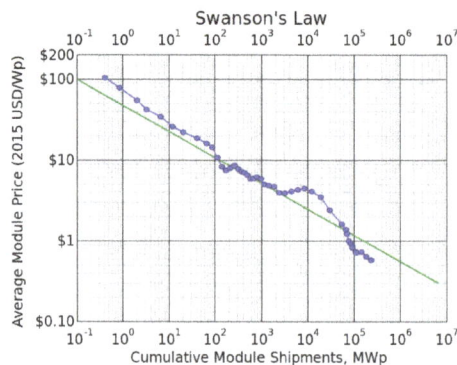

Swanson's law – the learning curve of solar PV

Adjusting for inflation, it cost $96 per watt for a solar module in the mid-1970s. Process improvements and a very large boost in production have brought that figure down 99%, to 68¢ per watt in 2016, according to data from Bloomberg New Energy Finance. Swanson's law is an observation similar to Moore's Law that states that solar cell prices fall 20% for every doubling of industry capacity. It was featured in an article in the British weekly newspaper The Economist in late 2012.

Further improvements reduced production cost to under $1 per watt, with wholesale costs well under $2. Balance of system costs were then higher than those of the panels. Large commercial arrays could be built, as of 2010, at below $3.40 a watt, fully commissioned.

As the semiconductor industry moved to ever-larger boules, older equipment became inexpensive. Cell sizes grew as equipment became available on the surplus market; ARCO Solar's original panels used cells 2 to 4 inches (50 to 100 mm) in diameter. Panels in the 1990s and early 2000s generally used 125 mm wafers; since 2008 almost all new panels use 156 mm cells. The widespread introduction of flat screen televisions in the late 1990s and early 2000s led to the wide availability of large, high-quality glass sheets to cover the panels.

During the 1990s, polysilicon ("poly") cells became increasingly popular. These cells offer less efficiency than their monosilicon ("mono") counterparts, but they are grown in large vats that reduce cost. By the mid-2000s, poly was dominant in the low-cost panel market, but more recently the mono returned to widespread use.

Manufacturers of wafer-based cells responded to high silicon prices in 2004–2008 with rapid reductions in silicon consumption. In 2008, according to Jef Poortmans, director of IMEC's organic and solar department, current cells use 8–9 grams (0.28–0.32 oz) of silicon per watt of power generation, with wafer thicknesses in the neighborhood of 200 microns. Crystalline silicon panels dominate worldwide markets and are mostly manufactured in China and Taiwan. By late 2011, a drop in European demand dropped prices for crystalline solar modules to about $1.09 per watt down sharply from 2010. Prices continued to fall in 2012, reaching $0.62/watt by 4Q2012.

Global installed PV capacity reached at least 177 gigawatts in 2014, enough to supply 1 percent of the world's total electricity consumption. Solar PV is growing fastest in Asia, with China and Japan currently accounting for half of worldwide deployment.

Subsidies and Grid Parity

Solar-specific feed-in tariffs vary by country and within countries. Such tariffs encourage the development of solar power projects. Widespread grid parity, the point at which photovoltaic electricity is equal to or cheaper than grid power without subsidies, likely requires advances on all three fronts. Proponents of solar hope to achieve grid parity first in areas with abundant sun and high electricity costs such as in California and Japan. In 2007 BP claimed grid parity for Hawaii and other islands that otherwise use diesel fuel to produce electricity. George W. Bush set 2015 as the date for grid parity in the US. The Photovoltaic Association reported in 2012 that Australia had reached grid parity (ignoring feed in tariffs).

The price of solar panels fell steadily for 40 years, interrupted in 2004 when high subsidies in Germany drastically increased demand there and greatly increased the price of purified silicon (which is used in computer chips as well as solar panels). The recession of 2008 and the onset of Chinese manufacturing caused prices to resume their decline. In the four years after January 2008 prices for solar modules in Germany dropped from €3 to €1 per peak watt. During that same time production capacity surged with an annual growth of more than 50%. China increased market share from 8% in 2008 to over 55% in the last quarter of 2010. In December 2012 the price of Chinese solar panels had dropped to $0.60/Wp (crystalline modules). (The abbreviation Wp stands for watt peak capacity, or the maximum capacity under optimal conditions.)

As of the end of 2016, it was reported that spot prices for assembled solar *panels* (not cells) had fallen to a record-low of US$0.36/Wp. It was estimated that the world's largest producer, Trina Solar Ltd. of China, was probably selling at a loss. The second largest supplier, Canadian Solar Inc., had reported costs of US$0.37/Wp in the third quarter of 2016, having dropped $0.02 from the previous quarter, and hence was probably still at least breaking even. Many producers expected costs would drop to the vicinity of $0.30 by the end of 2017. It was also reported that new solar installations were cheaper that coal-based thermal power plants in some regions of the world, and this was expected to be the case in most of the world within a decade.

Theory

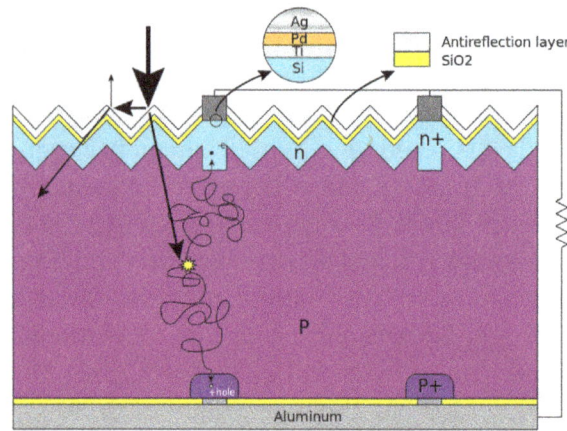

Working mechanism of a solar cell

The solar cell works in several steps:

- Photons in sunlight hit the solar panel and are absorbed by semiconducting materials, such as silicon.

- Electrons are excited from their current molecular/atomic orbital. Once excited an electron can either dissipate the energy as heat and return to its orbital or travel through the cell until it reaches an electrode. Current flows through the material to cancel the potential and this electricity is captured. The chemical bonds of the material are vital for this process to work, and usually silicon is used in two layers, one layer being doped with boron, the other phosphorus. These layers have different chemical electric charges and subsequently both drive and direct the current of electrons.

- An array of solar cells converts solar energy into a usable amount of direct current (DC) electricity.

- An inverter can convert the power to alternating current (AC).

The most commonly known solar cell is configured as a large-area p–n junction made from silicon.

The illuminated side of a solar cell must have a transparent conducting film to collect the current produced. Films such as indium tin oxide, conducting polymers or conducting nanowire networks are used.

Efficiency

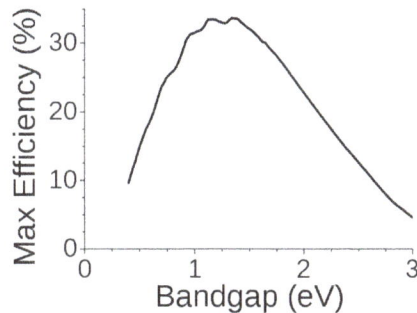

The Shockley-Queisser limit for the theoretical maximum efficiency of a solar cell. Semiconductors with band gap between 1 and 1.5eV, or near-infrared light, have the greatest potential to form an efficient single-junction cell. (The efficiency "limit" shown here can be exceeded by multijunction solar cells.)

Solar cell efficiency may be broken down into reflectance efficiency, thermodynamic efficiency, charge carrier separation efficiency and conductive efficiency. The overall efficiency is the product of these individual metrics.

A solar cell has a voltage dependent efficiency curve, temperature coefficients, and allowable shadow angles.

Due to the difficulty in measuring these parameters directly, other parameters are substituted: thermodynamic efficiency, quantum efficiency, integrated quantum efficiency, V_{oc} ratio, and fill factor. Reflectance losses are a portion of quantum efficiency under "external quantum efficiency". Recombination losses make up another portion of quantum efficiency, V_{oc} ratio, and fill factor. Resistive losses are predominantly categorized under fill factor, but also make up minor portions of quantum efficiency, V_{oc} ratio.

The fill factor is the ratio of the actual maximum obtainable power to the product of the open circuit voltage and short circuit current. This is a key parameter in evaluating performance. In 2009, typical commercial solar cells had a fill factor > 0.70. Grade B cells were usually between 0.4 and 0.7. Cells with a high fill factor have a low equivalent series resistance and a high equivalent shunt resistance, so less of the current produced by the cell is dissipated in internal losses.

Single p–n junction crystalline silicon devices are now approaching the theoretical limiting power efficiency of 33.16%, noted as the Shockley–Queisser limit in 1961. In the extreme, with an infinite number of layers, the corresponding limit is 86% using concentrated sunlight.

In 2014, three companies broke the record of 25.6% for a silicon solar cell. Panasonic's was the most efficient. The company moved the front contacts to the rear of the panel, eliminating shaded areas. In addition they applied thin silicon films to the (high quality silicon) wafer's front and back to eliminate defects at or near the wafer surface.

In 2015, a 4-junction GaInP/GaAs//GaInAsP/GaInAs solar cell achieved a new laboratory record efficiency of 46.1 percent (concentration ratio of sunlight = 312) in a French-German collaboration between the Fraunhofer Institute for Solar Energy Systems (Fraunhofer ISE), CEA-LETI and SOITEC.

In September 2015, Fraunhofer ISE announced the achievement of an efficiency above 20% for epitaxial wafer cells. The work on optimizing the atmospheric-pressure chemical vapor deposition (APCVD) in-line production chain was done in collaboration with NexWafe GmbH, a company spun off from Fraunhofer ISE to commercialize production.

For triple-junction thin-film solar cells, the world record is 13.6%, set in June 2015.

In 2016, researchers at Fraunhofer ISE announced a GaInP/GaAs/Si triple-junction solar cell with two terminals reaching 30.2% efficiency without concentration.

In 2017, researchers at National Renewable Energy Laboratory (NREL) in Golden, Colorado set the world record for dual-junction, monolithic two-terminal solar cell reaching 32.6% efficiency under one-sun.

Materials

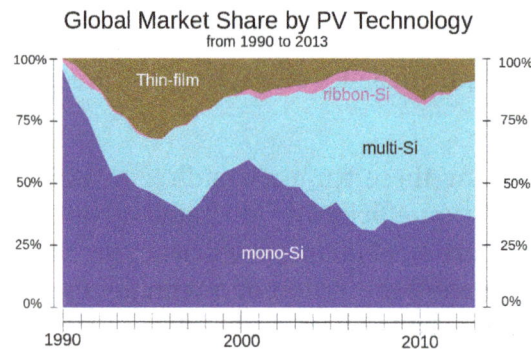

Global market-share in terms of annual production by PV technology since 1990.

Solar cells are typically named after the semiconducting material they are made of. These materials must have certain characteristics in order to absorb sunlight. Some cells are designed to handle sunlight that reaches the Earth's surface, while others are optimized for use in space. Solar cells can be made of only one single layer of light-absorbing material (single-junction) or use multiple physical configurations (multi-junctions) to take advantage of various absorption and charge separation mechanisms.

Solar cells can be classified into first, second and third generation cells. The first generation cells—also called conventional, traditional or wafer-based cells—are made of crystalline silicon, the commercially predominant PV technology, that includes materials such as polysilicon and monocrystalline silicon. Second generation cells are thin film solar cells, that include amorphous silicon, CdTe and CIGS cells and are commercially significant in utility-scale photovoltaic power stations, building integrated photovoltaics or in small stand-alone power system. The third generation of solar cells includes a number of thin-film technologies often described as emerging photovoltaics—most of them have not yet been commercially applied and are still in the research or development phase. Many use organic materials, often organometallic compounds as well as inorganic substances. Despite the fact that their efficiencies had been low and the stability of the absorber material was often too short for commercial applications, there is a lot of research invested into these technologies as they promise to achieve the goal of producing low-cost, high-efficiency solar cells.

Crystalline Silicon

By far, the most prevalent bulk material for solar cells is crystalline silicon (c-Si), also known as "solar grade silicon". Bulk silicon is separated into multiple categories according to crystallinity and crystal size in the resulting ingot, ribbon or wafer. These cells are entirely based around the concept of a p-n junction. Solar cells made of c-Si are made from wafers between 160 and 240 micrometers thick.

Monocrystalline Silicon

Monocrystalline silicon (mono-Si) solar cells are more efficient and more expensive than most other types of cells. The corners of the cells look clipped, like an octagon, because the wafer material is cut from cylindrical ingots, that are typically grown by the Czochralski process. Solar panels using mono-Si cells display a distinctive pattern of small white diamonds.

Epitaxial Silicon Development

Epitaxial wafers of crystalline silicon can be grown on a monocrystalline silicon "seed" wafer by chemical vapor deposition (CVD), and then detached as self-supporting wafers of some standard thickness (e.g., 250 µm) that can be manipulated by hand, and directly substituted for wafer cells cut from monocrystalline silicon ingots. Solar cells made with this "kerfless" technique can have efficiencies approaching those of wafer-cut cells, but at appreciably lower cost if the CVD can be done at atmospheric pressure in a high-throughput inline process. The surface of epitaxial wafers may be textured to enhance light absorption.

In June 2015, it was reported that heterojunction solar cells grown epitaxially on n-type monocrystalline silicon wafers had reached an efficiency of 22.5% over a total cell area of 243.4 cm^2.

Polycrystalline Silicon

Polycrystalline silicon, or multicrystalline silicon (multi-Si) cells are made from cast square ingots—large blocks of molten silicon carefully cooled and solidified. They consist of small crystals giving the material its typical metal flake effect. Polysilicon cells are the most common type used in photovoltaics and are less expensive, but also less efficient, than those made from monocrystalline silicon.

Ribbon Silicon

Ribbon silicon is a type of polycrystalline silicon—it is formed by drawing flat thin films from molten silicon and results in a polycrystalline structure. These cells are cheaper to make than multi-Si, due to a great reduction in silicon waste, as this approach does not require sawing from ingots. However, they are also less efficient.

Mono-like-multi Silicon (MLM)

This form was developed in the 2000s and introduced commercially around 2009. Also called cast-mono, this design uses polycrystalline casting chambers with small "seeds" of mono material. The result is a bulk mono-like material that is polycrystalline around the outsides. When sliced for processing, the inner sections are high-efficiency mono-like cells (but square instead of "clipped"), while the outer edges are sold as conventional poly. This production method results in mono-like cells at poly-like prices.

Thin Film

Thin-film technologies reduce the amount of active material in a cell. Most designs sandwich active material between two panes of glass. Since silicon solar panels only use one pane of glass, thin film panels are approximately twice as heavy as crystalline silicon panels, although they have a smaller ecological impact (determined from life cycle analysis).

Cadmium Telluride

Cadmium telluride is the only thin film material so far to rival crystalline silicon in cost/watt. However cadmium is highly toxic and tellurium (anion: "telluride") supplies are limited. The cadmium present in the cells would be toxic if released. However, release is impossible during normal operation of the cells and is unlikely during fires in residential roofs. A square meter of CdTe contains approximately the same amount of Cd as a single C cell nickel-cadmium battery, in a more stable and less soluble form.

Copper Indium Gallium Selenide

Copper indium gallium selenide (CIGS) is a direct band gap material. It has the highest efficiency (~20%) among all commercially significant thin film materials. Traditional methods of fabrication involve vacuum processes including co-evaporation and sputtering. Recent developments at IBM and Nanosolar attempt to lower the cost by using non-vacuum solution processes.

Silicon Thin Film

Silicon thin-film cells are mainly deposited by chemical vapor deposition (typically plasma-enhanced, PE-CVD) from silane gas and hydrogen gas. Depending on the deposition parameters, this can yield amorphous silicon (a-Si or a-Si:H), protocrystalline silicon or nanocrystalline silicon (nc-Si or nc-Si:H), also called microcrystalline silicon.

Amorphous silicon is the most well-developed thin film technology to-date. An amorphous silicon (a-Si) solar cell is made of non-crystalline or microcrystalline silicon. Amorphous silicon has a higher bandgap (1.7 eV) than crystalline silicon (c-Si) (1.1 eV), which means it absorbs the visible part of the solar spectrum more strongly than the higher power density infrared portion of the spectrum. The production of a-Si thin film solar cells uses glass as a substrate and deposits a very thin layer of silicon by plasma-enhanced chemical vapor deposition (PECVD).

Protocrystalline silicon with a low volume fraction of nanocrystalline silicon is optimal for high open circuit voltage. Nc-Si has about the same bandgap as c-Si and nc-Si and a-Si can advantageously be combined in thin layers, creating a layered cell called a tandem cell. The top cell in a-Si absorbs the visible light and leaves the infrared part of the spectrum for the bottom cell in nc-Si.

Gallium Arsenide Thin Film

The semiconductor material Gallium arsenide (GaAs) is also used for single-crystalline thin film solar cells. Although GaAs cells are very expensive, they hold the world's record in efficiency for a single-junction solar cell at 28.8%. GaAs is more commonly used in multijunction photovoltaic cells for concentrated photovoltaics (CPV, HCPV) and for solar panels on spacecrafts, as the industry favours efficiency over cost for space-based solar power.

Multijunction Cells

Dawn's 10 kW triple-junction gallium arsenide solar array at full extension.

Multi-junction cells consist of multiple thin films, each essentially a solar cell grown on top of another, typically using metalorganic vapour phase epitaxy. Each layers has a different band gap energy to allow it to absorb electromagnetic radiation over a different portion of the spectrum. Multi-junction cells were originally developed for special applications such as satellites and space exploration, but are now used increasingly in terrestrial concentrator photovoltaics (CPV), an emerging technology that uses lenses and curved mirrors to concentrate sunlight onto small, highly efficient multi-junction solar cells. By concentrating sunlight up to a thousand times, *High concentrated photovoltaics (HCPV)* has the potential to outcompete conventional solar PV in the future.

Tandem solar cells based on monolithic, series connected, gallium indium phosphide (GaInP), gallium arsenide (GaAs), and germanium (Ge) p–n junctions, are increasing sales, despite cost pressures. Between December 2006 and December 2007, the cost of 4N gallium metal rose from about $350 per kg to $680 per kg. Additionally, germanium metal prices have risen substantially to $1000–1200 per kg this year. Those materials include gallium (4N, 6N and 7N Ga), arsenic (4N, 6N and 7N) and germanium, pyrolitic boron nitride (pBN) crucibles for growing crystals, and boron oxide, these products are critical to the entire substrate manufacturing industry.

A triple-junction cell, for example, may consist of the semiconductors: GaAs, Ge, and GaInP 2. Triple-junction GaAs solar cells were used as the power source of the Dutch four-time World Solar Challenge winners Nuna in 2003, 2005 and 2007 and by the Dutch solar cars Solutra (2005), Twente One (2007) and 21Revolution (2009). GaAs based multi-junction devices are the most efficient solar cells to date. On 15 October 2012, triple junction metamorphic cells reached a record high of 44%.

In 2016, a new approach was described for producing hybrid photovoltaic wafers combining the high efficiency of III-V multi-junction solar cells with the economies and wealth of experience associated with silicon. The technical complications involved in growing the III-V material on silicon at the required high temperatures, a subject of study for some 30 years, are avoided by epitaxial growth of silicon on GaAs at low temperature by plasma-enhanced chemical vapor deposition (PECVD).

Research in Solar Cells

Perovskite Solar Cells

Perovskite solar cells are solar cells that include a perovskite-structured material as the active layer. Most commonly, this is a solution-processed hybrid organic-inorganic tin or lead halide based material. Efficiencies have increased from below 5% at their first usage in 2009 to over 20% in 2014, making them a very rapidly advancing technology and a hot topic in the solar cell field. Perovskite solar cells are also forecast to be extremely cheap to scale up, making them a very attractive option for commercialisation.

Liquid Inks

In 2014, researchers at California NanoSystems Institute discovered using kesterite and perovskite improved electric power conversion efficiency for solar cells.

Upconversion and Downconversion

Photon upconversion is the process of using two low-energy (*e.g.,* infrared) photons to produce one higher energy photon; downconversion is the process of using one high energy photon (*e.g.,,* ultraviolet) to produce two lower energy photons. Either of these techniques could be used to produce higher efficiency solar cells by allowing solar photons to be more efficiently used. The difficulty, however, is that the conversion efficiency of existing phosphors exhibiting up- or down-conversion is low, and is typically narrow band.

One upconversion technique is to incorporate lanthanide-doped materials (Er^{3+}, Yb^{3+}, Ho^{3+} or a combination), taking advantage of their luminescence to convert infrared radiation to visible light. Upconversion process occurs when two infrared photons are absorbed by rare-earth ions to generate a (high-energy) absorbable photon. As example, the energy transfer upconversion process (ETU), consists in successive transfer processes between excited ions in the near infrared. The upconverter material could be placed below the solar cell to absorb the infrared light that passes through the silicon. Useful ions are most commonly found in the trivalent state. Er^+ ions have been the most used. Er^{3+} ions absorb solar radiation around 1.54 µm. Two Er^{3+} ions that have absorbed this radiation can interact with each other through an upconversion process. The excited ion emits light above the Si bandgap that is absorbed by the solar cell and creates an additional electron–hole pair that can generate current. However, the increased efficiency was small. In addition, fluoroindate glasses have low phonon energy and have been proposed as suitable matrix doped with Ho^{3+} ions.

Light-absorbing Dyes

Dye-sensitized solar cells (DSSCs) are made of low-cost materials and do not need elaborate manufacturing equipment, so they can be made in a DIY fashion. In bulk it should be significantly less expensive than older solid-state cell designs. DSSC's can be engineered into flexible sheets and although its conversion efficiency is less than the best thin film cells, its price/performance ratio may be high enough to allow them to compete with fossil fuel electrical generation.

Typically a ruthenium metalorganic dye (Ru-centered) is used as a monolayer of light-absorbing material. The dye-sensitized solar cell depends on a mesoporous layer of nanoparticulate titanium dioxide to greatly amplify the surface area (200–300 m²/g TiO_2, as compared to approximately 10 m²/g of flat single crystal). The photogenerated electrons from the light absorbing dye are passed on to the n-type TiO_2 and the holes are absorbed by an electrolyte on the other side of the dye. The circuit is completed by a redox couple in the electrolyte, which can be liquid or solid. This type of cell allows more flexible use of materials and is typically manufactured by screen printing or ultrasonic nozzles, with the potential for lower processing costs than those used for bulk solar cells. However, the dyes in these cells also suffer from degradation under heat and UV light and the cell casing is difficult to seal due to the solvents used in assembly. The first commercial shipment of DSSC solar modules occurred in July 2009 from G24i Innovations.

Quantum Dots

Quantum dot solar cells (QDSCs) are based on the Gratzel cell, or dye-sensitized solar cell architecture, but employ low band gap semiconductor nanoparticles, fabricated with crystallite sizes small enough to form quantum dots (such as CdS, CdSe, Sb_2S3, PbS, etc.), instead of organic or organometallic dyes as light absorbers. QD's size quantization allows for the band gap to be tuned by simply changing particle size. They also have high extinction coefficients and have shown the possibility of multiple exciton generation.

In a QDSC, a mesoporous layer of titanium dioxide nanoparticles forms the backbone of the cell, much like in a DSSC. This TiO_2 layer can then be made photoactive by coating with semiconductor quantum dots using chemical bath deposition, electrophoretic deposition or successive ionic layer adsorption and reaction. The electrical circuit is then completed through the use of a liquid or solid redox couple. The efficiency of QDSCs has increased to over 5% shown for both liquid-junction and solid state cells. In an effort to decrease production costs, the Prashant Kamat research group demonstrated a solar paint made with TiO_2 and CdSe that can be applied using a one-step method to any conductive surface with efficiencies over 1%.

Organic/Polymer Solar Cells

Organic solar cells and polymer solar cells are built from thin films (typically 100 nm) of organic semiconductors including polymers, such as polyphenylene vinylene and small-molecule compounds like copper phthalocyanine (a blue or green organic pigment) and carbon fullerenes and fullerene derivatives such as PCBM.

They can be processed from liquid solution, offering the possibility of a simple roll-to-roll printing process, potentially leading to inexpensive, large-scale production. In addition, these cells could be beneficial for some applications where mechanical flexibility and disposability are important. Current cell efficiencies are, however, very low, and practical devices are essentially non-existent.

Energy conversion efficiencies achieved to date using conductive polymers are very low compared to inorganic materials. However, Konarka Power Plastic reached efficiency of 8.3% and organic tandem cells in 2012 reached 11.1%.

The active region of an organic device consists of two materials, one electron donor and one electron acceptor. When a photon is converted into an electron hole pair, typically in the donor material, the charges tend to remain bound in the form of an exciton, separating when the exciton diffuses to the donor-acceptor interface, unlike most other solar cell types. The short exciton diffusion lengths of most polymer systems tend to limit the efficiency of such devices. Nanostructured interfaces, sometimes in the form of bulk heterojunctions, can improve performance.

In 2011, MIT and Michigan State researchers developed solar cells with a power efficiency close to 2% with a transparency to the human eye greater than 65%, achieved by selectively absorbing the ultraviolet and near-infrared parts of the spectrum with small-molecule compounds. Researchers at UCLA more recently developed an analogous polymer solar cell, following the same approach, that is 70% transparent and has a 4% power conversion efficiency. These lightweight, flexible cells can be produced in bulk at a low cost and could be used to create power generating windows.

In 2013, researchers announced polymer cells with some 3% efficiency. They used block copolymers, self-assembling organic materials that arrange themselves into distinct layers. The research focused on P3HT-b-PFTBT that separates into bands some 16 nanometers wide.

Adaptive Cells

Adaptive cells change their absorption/reflection characteristics depending to respond to environmental conditions. An adaptive material responds to the intensity and angle of incident light. At the part of the cell where the light is most intense, the cell surface changes from reflective to adaptive, allowing the light to penetrate the cell. The other parts of the cell remain reflective increasing the retention of the absorbed light within the cell.

In 2014 a system was developed that combined an adaptive surface with a glass substrate that redirect the absorbed to a light absorber on the edges of the sheet. The system also includes an array of fixed lenses/mirrors to concentrate light onto the adaptive surface. As the day continues, the concentrated light moves along the surface of the cell. That surface switches from reflective to adaptive when the light is most concentrated and back to reflective after the light moves along.

Surface Texturing

For the past years, researchers have been trying to reduce the price of solar cells while maximizing efficiency. Thin-film solar cell is a cost-effective second generation solar cell with much reduced thickness at the expense of light absorption efficiency. Efforts to maximize light absorption efficiency with reduced thickness have been made. Surface texturing is one of techniques used to reduce optical losses to maximize light absorbed. Currently, surface texturing techniques on silicon photovoltaics are drawing much attention. Surface texturing could be done in multiple ways. Etching single crystalline silicon substrate can produce randomly distributed square based pyramids on the surface using anisotropic etchants. Recent studies show that c-Si wafers could be etched down to form nano-scale inverted pyramids. Multicrystalline silicon solar cells, due to poorer crystallographic quality, are less effective than single crystal solar cells, but mc-Si solar cells are still being used widely due to less manufacturing difficulties. It is reported that multicrystalline solar cells can be surface-textured to yield solar energy conversion efficiency comparable to that of monocrystalline silicon cells, through isotropic etching or photolithography techniques. Incident light rays onto a textured surface do not reflect back out to the air as opposed to rays onto a flat surface. Rather some light rays are bounced back onto the other surface again due to the geometry of the surface. This process significantly improves light to electricity conversion efficiency, due to increased light absorption. This texture effect as well as the interaction with other interfaces in the PV module is a challenging optical simulation task. A particularly efficient method for modeling and optimization is the OPTOS formalism. In 2012, researchers at MIT reported that c-Si films textured with nanoscale inverted pyramids could achieve light absorption comparable to 30 times thicker planar c-Si. In combination with anti-reflective coating, surface texturing technique can effectively trap light rays within a thin film silicon solar cell. Consequently, required thickness for solar cells decreases with the increased absorption of light rays.

Encapsulation

Solar cells are commonly encapsulated in a transparent polymeric resin to protect the delicate solar cell regions for coming into contact with moisture, dirt, ice, and other conditions expected either during operation or when used outdoors. The encapsulants are commonly made from polyvinyl acetate or glass. Most encapsulants are uniform in structure and composition, which increases light collection owing to light trapping from total internal reflection of light within the resin. Research has been conducted into structuring the encapsulant to provide further collection of light. Such encapsulants have included roughened glass surfaces, diffractive elements, prism arrays, air prisms, v-grooves, diffuse elements, as well as multi-directional waveguide arrays. Prism arrays show an overall 5% increase in the total solar energy conversion. Active coatings that convert infrared light into visible light have shown a 30% increase.

Manufacture

Solar cells share some of the same processing and manufacturing techniques as other semiconductor devices. However, the stringent requirements for cleanliness and quality control of semiconductor fabrication are more relaxed for solar cells, lowering costs.

Polycrystalline silicon wafers are made by wire-sawing block-cast silicon ingots into 180 to 350 micrometer wafers. The wafers are usually lightly p-type-doped. A surface diffusion of n-type dopants is performed on the front side of the wafer. This forms a p–n junction a few hundred nanometers below the surface.

Anti-reflection coatings are then typically applied to increase the amount of light coupled into the solar cell. Silicon nitride has gradually replaced titanium dioxide as the preferred material, because of its excellent surface passivation qualities. It prevents carrier recombination at the cell surface. A layer several hundred nanometers thick is applied using PECVD. Some solar cells have textured front surfaces that, like anti-reflection coatings, increase the amount of light reaching the wafer. Such surfaces were first applied to single-crystal silicon, followed by multicrystalline silicon somewhat later.

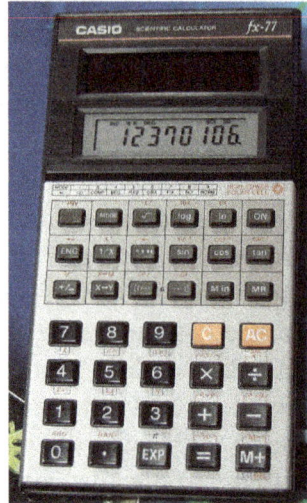

Early solar-powered calculator

A full area metal contact is made on the back surface, and a grid-like metal contact made up of fine "fingers" and larger "bus bars" are screen-printed onto the front surface using a silver paste. This is an evolution of the so-called "wet" process for applying electrodes, first described in a US patent filed in 1981 by Bayer AG. The rear contact is formed by screen-printing a metal paste, typically aluminium. Usually this contact covers the entire rear, though some designs employ a grid pattern. The paste is then fired at several hundred degrees Celsius to form metal electrodes in ohmic contact with the silicon. Some companies use an additional electro-plating step to increase efficiency. After the metal contacts are made, the solar cells are interconnected by flat wires or metal ribbons, and assembled into modules or "solar panels". Solar panels have a sheet of tempered glass on the front, and a polymer encapsulation on the back.

Manufacturers and Certification

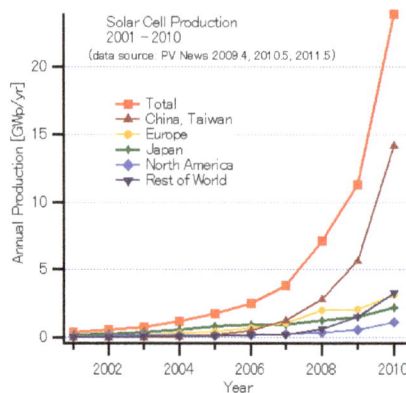

Solar cell production by region

National Renewable Energy Laboratory tests and validates solar technologies. Three reliable groups certify solar equipment: UL and IEEE (both U.S. standards) and IEC.

Solar cells are manufactured in volume in Japan, Germany, China, Taiwan, Malaysia and the United States, whereas Europe, China, the U.S., and Japan have dominated (94% or more as of 2013) in installed systems. Other nations are acquiring significant solar cell production capacity.

Global PV cell/module production increased by 10% in 2012 despite a 9% decline in solar energy investments according to the annual "PV Status Report" released by the European Commission's Joint Research Centre. Between 2009 and 2013 cell production has quadrupled.

China

Due to heavy government investment, China has become the dominant force in solar cell manufacturing. Chinese companies produced solar cells/modules with a capacity of ~23 GW in 2013 (60% of global production).

Malaysia

In 2014, Malaysia was the world's third largest manufacturer of photovoltaics equipment, behind China and the European Union.

United States

Solar cell production in the U.S. has suffered due to the global financial crisis, but recovered partly due to the falling price of quality silicon.

Solar Charger

A solar charger employs solar energy to supply electricity to devices or charge batteries. They are generally portable.

Solar chargers can charge lead acid or Ni-Cd battery banks up to 48 V and hundreds of ampere-hours (up to 4000 Ah) capacity. Such type of solar charger setups generally use an intelligent charge controller. A series of solar cells are installed in a stationary location (ie: rooftops of homes, base-station locations on the ground etc.) and can be connected to a battery bank to store energy for off-peak usage. They can also be used in addition to mains-supply chargers for energy saving during the daytime.

Suntactics sCharger-5

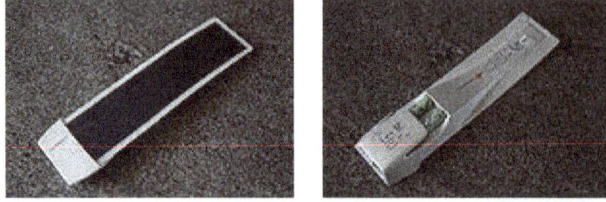

Front and back views of a small, portable solar charger with two AAA Batteries and USB output.

Solar charger integrated into a backpack

Most portable chargers can obtain energy from the sun only. Some, including the Kinesis K3, and GeNNex Solar Cell 2 can work either way (recharged by the sun or plugged into a wall plug to charge up). Examples of solar chargers in popular use include:

- Small portable models designed to charge a range of different mobile phones, cell phones, iPods or other portable audio equipment.

- Fold out models designed to sit on the dashboard of an automobile and plug into the cigar/12v lighter socket to keep the battery topped up while the vehicle is not in use.

- Flashlights/torches, often combined with a secondary means of charging, such as a kinetic (hand crank generator) charging system.

- Public solar chargers permanently installed in public places, such as parks, squares and streets, which anyone can use for free.

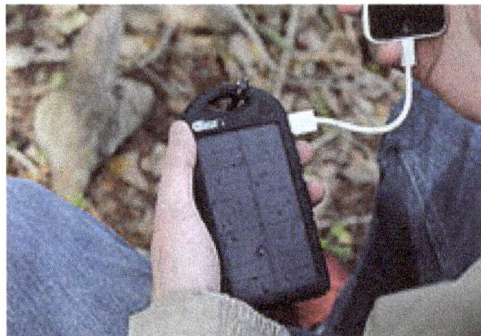

Creative Edge Solar-5, 1.2w solar panel with 5000mAh internal battery

Voltage Regulator

A solar panel can produce a range of charging voltages depending upon sunlight intensity, so a voltage regulator must be included in the charging circuit so as to not over-charge (over voltage) a device such as a 12 volt car battery.

Solar Chargers on the Market

Portable solar chargers are used to charge cell phones and other small electronic devices on the go. Chargers on the market today use various types of solar panels, ranging from thin film panels with efficiencies from 7-15% (amorphous silicon around 7%, CIGS closer to 15%), to the slightly more efficient monocrystalline panels which offer efficiencies up to 18%.

The other type of portable solar chargers are those with wheels which enable them to be transported from one place to another and be used by a lot of people. They are semi-public, considering the fact that are used publicly but not permanently installed. A good example of this kind of portable solar charger is the Strawberry Mini device.

Public solar charger Strawberry Tree Black in Tašmajdan Park in Belgrade

The solar charger industry has been plagued by companies mass-producing low efficiency solar chargers that don't meet the consumer's expectations. This in turn has made it hard for new solar charger companies to gain the trust of consumers. Solar companies are starting to offer high-efficiency solar chargers. When it comes to permanently installed public solar chargers, Strawberry energy Company from Serbia has invented and developed the first public solar charger for mobile devices, Strawberry Tree. Due to a built in rechargeable battery which stores energy, it can function without sunshine or at night. Other companies such as Voltaic Systems, Poweradd and others have started to push better products onto the market as well.

Rural African villagers holding portable solar charger

Portable solar power is being utilized in developing countries to power lighting as opposed to utilizing kerosene lamps which are responsible for respiratory infections, lung and throat cancers, serious eye infections, cataracts as well as low birth weights. Solar power provides an opportunity for rural areas to "leapfrog" traditional grid infrastructure and move directly to distributed energy solutions.

Some solar chargers also have an on-board battery which is charged by the solar panel when not charging anything else. This allows the user to be able to use the solar energy stored in the battery to charge their electronic devices at night or when indoors.

Solar chargers can also be rollable or flexible and are manufactured using thin film PV technology. Rollable solar chargers may include Li-ion batteries.

Currently, foldable solar panels are coming down in price to the point that almost anyone can deploy one while at the beach, biking, hiking, or at any outdoor location and charge their cellphone, tablet, computer etc. As advances in the technology continue to be made, new products will be able to be expand into third world countries faster than ever before. Recently, billionaire Elon Musk unveiled a new battery system that may allow off-grid systems that rely solely on solar power to recharge them to be deployed anywhere on the planet, thus possibly ending the need for strictly grid-based energy systems.

Theory of Solar Cells

The theory of solar cells explains the process by which light energy in photons is converted into electric current when the photons strike a suitable semiconductor device. The theoretical studies are of practical use because they predict the fundamental limits of solar cell, and give guidance on the phenomena that contribute to losses and solar cell efficiency.

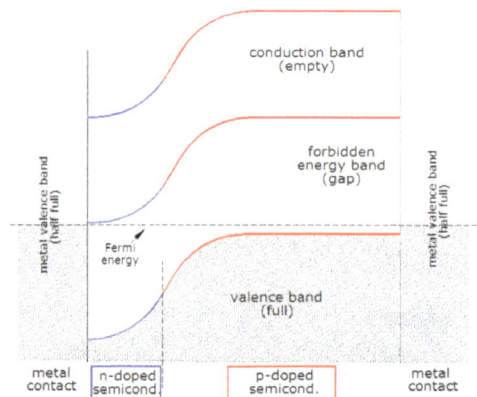

Band diagram of a solar cell, corresponding to very low current (horizontal Fermi level), very low voltage (metal valence bands at same height), and therefore very low illumination.

Simple Explanation

1. Photons in sunlight hit the solar panel and are absorbed by semi-conducting materials.

2. Electrons (negatively charged) are knocked loose from their atoms as they are excited. Due

to their special structure and the materials in solar cells, the electrons are only allowed to move in a single direction. The electronic structure of the materials is very important for the process to work, and often silicon incorporating small amounts of boron or phosphorus is used in different layers.

3. An array of solar cells converts solar energy into a usable amount of direct current (DC) electricity.

Photogeneration of Charge Carriers

When a photon hits a piece of silicon, one of three things can happen:

1. The photon can pass straight through the silicon — this (generally) happens for lower energy photons.

2. The photon can reflect off the surface.

3. The photon can be absorbed by the silicon if the photon energy is higher than the silicon band gap value. This generates an electron-hole pair and sometimes heat depending on the band structure.

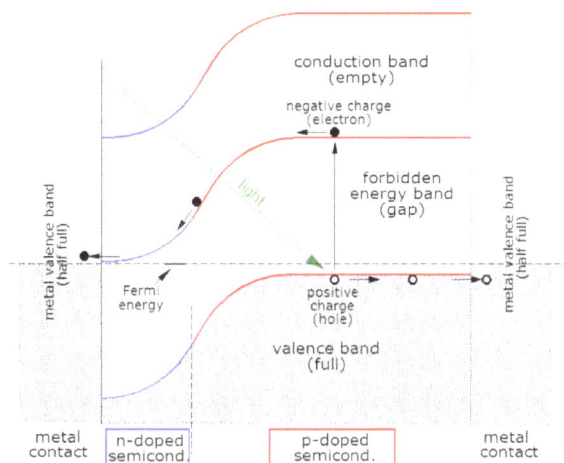

Band diagram of a silicon solar cell, corresponding to very low current (horizontal Fermi level), very low voltage (metal valence bands at same height), and therefore very low illumination.

When a photon is absorbed, its energy is given to an electron in the crystal lattice. Usually this electron is in the valence band and is tightly bound in covalent bonds with neighboring atoms, and therefore unable to move far. The energy given to the electron by the photon "excites" it into the conduction band where it is free to move around within the semiconductor. The network of covalent bonds that the electron was previously a part of now has one fewer electron. This is known as a hole. The presence of a missing covalent bond allows the bonded electrons of neighboring atoms to move into the "hole," leaving another hole behind, thus propagating holes throughout the lattice. It can be said that photons absorbed in the semiconductor create electron-hole pairs.

A photon only needs to have energy greater than that of the band gap in order to excite an electron from the valence band into the conduction band. However, the solar frequency spectrum approximates a black body spectrum at about 5,800 K, and as such, much of the solar radiation reaching the Earth is composed of photons with energies greater than the band gap of silicon. These higher energy photons will be absorbed by the solar cell, but the difference in energy between these photons and the silicon band gap is converted into heat (via lattice vibrations — called phonons) rather than into usable electrical energy. The photovoltaic effect can also occur when two photons are absorbed simultaneously in a process called two-photon photovoltaic effect. However, high optical intensities are required for this nonlinear process.

The p-n Junction

The most commonly known solar cell is configured as a large-area p-n junction made from silicon. As a simplification, one can imagine bringing a layer of n-type silicon into direct contact with a layer of p-type silicon. In practice, p-n junctions of silicon solar cells are not made in this way, but rather by diffusing an n-type dopant into one side of a p-type wafer (or vice versa).

If a piece of p-type silicon is placed in close contact with a piece of n-type silicon, then a diffusion of electrons occurs from the region of high electron concentration (the n-type side of the junction) into the region of low electron concentration (p-type side of the junction). When the electrons diffuse across the p-n junction, they recombine with holes on the p-type side. However (in the absence of an external circuit) this diffusion of carriers does not go on indefinitely because charges build up on either side of the junction and create an electric field. The electric field promotes charge flow, known as drift current, that opposes and eventually balances out the diffusion of electrons and holes. This region where electrons and holes have diffused across the junction is called the depletion region because it contains practically no mobile charge carriers. It is also known as the *space charge region*, although space charge extends a bit further in both directions than the depletion region.

Charge Carrier Separation

There are two causes of charge carrier motion and separation in a solar cell:

1. drift of carriers, driven by the electric field, with electrons being pushed one way and holes the other way.

2. diffusion of carriers from zones of higher carrier concentration to zones of lower carrier concentration (following a gradient of electrochemical potential).

These two "forces" may work one against the other at any given point in the cell. For instance, an electron moving through the junction from the p region to the n region is being pushed by the electric field against the concentration gradient. The same goes for a hole moving in the opposite direction.

It is easiest to understand how a current is generated when considering electron-hole pairs that are created in the depletion zone, which is where there is a strong electric field. The electron is pushed by this field toward the n side and the hole toward the p side. (This is opposite to the direction of current in a forward-biased diode, such as a light-emitting diode in operation.) When the pair is created outside the space charge zone, where the electric field is smaller, diffusion also acts to move the carriers, but the junction still plays a role by sweeping any electrons that reach it from the p side to the n side, and by sweeping any holes that reach it from the n side to the p side, thereby creating a concentration gradient outside the space charge zone.

In thick solar cells there is very little electric field in the active region outside the space charge zone, so the dominant mode of charge carrier separation is diffusion. In these cells the diffusion length of minority carriers (the length that photo-generated carriers can travel before they recombine) must be large compared to the cell thickness. In thin film cells (such as amorphous silicon), the diffusion length of minority carriers is usually very short due to the existence of defects, and the dominant charge separation is therefore drift, driven by the electrostatic field of the junction, which extends to the whole thickness of the cell.

Once the minority carrier enters the drift region, it is 'swept' across the junction and, at the other side of the junction, becomes a majority carrier. This reverse current is a generation current, fed both thermally and (if present) by the absorption of light. On the other hand, majority carriers are driven into the drift region by diffusion (resulting from the concentration gradient), which leads to the forward current; only the majority carriers with the highest energies (in the so-called Boltzmann tail; cf. Maxwell–Boltzmann statistics) can fully cross the drift region. Therefore, the carrier distribution in the whole device is governed by a dynamic equilibrium between reverse current and forward current.

Connection to an External Load

Ohmic metal-semiconductor contacts are made to both the n-type and p-type sides of the solar cell, and the electrodes connected to an external load. Electrons that are created on the n-type side, or created on the p-type side, "collected" by the junction and swept onto the n-type side, may travel through the wire, power the load, and continue through the wire until they reach the p-type semiconductor-metal contact. Here, they recombine with a hole that was either created as an electron-hole pair on the p-type side of the solar cell, or a hole that was swept across the junction from the n-type side after being created there.

The voltage measured is equal to the difference in the quasi Fermi levels of the majority carriers (electrons in the n-type portion and holes in the p-type portion) at the two terminals.

Equivalent Circuit of a Solar Cell

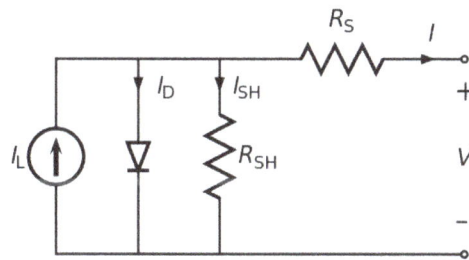

The equivalent circuit of a solar cell

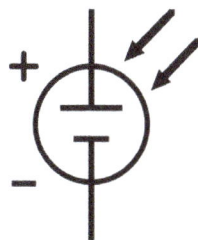

The schematic symbol of a solar cell

To understand the electronic behavior of a solar cell, it is useful to create a model which is electrically equivalent, and is based on discrete ideal electrical components whose behavior is well defined. An ideal solar cell may be modelled by a current source in parallel with a diode; in practice no solar cell is ideal, so a shunt resistance and a series resistance component are added to the model. The resulting equivalent circuit of a solar cell is shown. Also shown, is the schematic representation of a solar cell for use in circuit diagrams.

Characteristic Equation

From the equivalent circuit it is evident that the current produced by the solar cell is equal to that produced by the current source, minus that which flows through the diode, minus that which flows through the shunt resistor:

$$I = I_L - I_D - I_{SH}$$

where

- I = output current (ampere)

- I_L = photogenerated current (ampere)

- I_D = diode current (ampere)

- I_{SH} = shunt current (ampere).

The current through these elements is governed by the voltage across them:

$$V_j = V + IR_S$$

where

- V_j = voltage across both diode and resistor R_{SH} (volt)

- V = voltage across the output terminals (volt)

- I = output current (ampere)

- R_S = series resistance (Ω).

By the Shockley diode equation, the current diverted through the diode is:

$$I_D = I_0 \left\{ \exp\left[\frac{V_j}{nV_T} \right] - 1 \right\}$$

where

- I_o = reverse saturation current (ampere)

- n = diode ideality factor (1 for an ideal diode)

- q = elementary charge

- k = Boltzmann's constant

- T = absolute temperature

- $V_T = kT/q$, the thermal voltage. At 25 °C, $V_T \approx 0.0259$ volt.

By Ohm's law, the current diverted through the shunt resistor is:

$$I_{SH} = \frac{V_j}{R_{SH}}$$

where

- R_{SH} = shunt resistance (Ω).

Substituting these into the first equation produces the characteristic equation of a solar cell, which relates solar cell parameters to the output current and voltage:

$$I = I_L - I_0 \left\{ \exp\left[\frac{V + IR_S}{nV_T}\right] - 1 \right\} - \frac{V + IR_S}{R_{SH}}.$$

An alternative derivation produces an equation similar in appearance, but with V on the left-hand side. The two alternatives are identities; that is, they yield precisely the same results.

Since the parameters I_o, n, R_S, and R_{SH} cannot be measured directly, the most common application of the characteristic equation is nonlinear regression to extract the values of these parameters on the basis of their combined effect on solar cell behavior.

When R_S is not zero, the above equation does not give the current I directly, but it can then be solved using the Lambert W function:

$$I = \frac{(I_L + I_0) - V/R_{SH}}{1 + R_S/R_{SH}} - \frac{nV_T}{R_S} W\left(\frac{I_0 R_S}{nV_T(1 + R_S/R_{SH})} \exp\left(\frac{V}{nV_T}\left(1 - \frac{R_S}{R_S + R_{SH}}\right) + \frac{(I_L + I_0)R_S}{nV_T(1 + R_S/R_{SH})}\right)\right)$$

When an external load is used with the cell, its resistance can simply be added to R_S and V set to zero in order to find the current.

When R_{SH} is infinite there is a solution for V for any I less than $I_L + I_0$:

$$V = nV_T \ln\left(\frac{I_L - I}{I_0} + 1\right) - IR_S.$$

Otherwise one can solve for V using the Lambert W function:

$$V = (I_L + I_0)R_{SH} - I(R_S + R_{SH}) - nV_T W\left(\frac{I_0 R_{SH}}{nV_T} \exp\left(\frac{(I_L + I_0 - I)R_{SH}}{nV_T}\right)\right)$$

However, when R_{SH} is large it's better to solve the original equation numerically.

The general form of the solution is a curve with I decreasing as V increases. The slope at small or negative V (where the W function is near zero) approaches $-1/(R_S + R_{SH})$, whereas the slope at high V approaches $-1/R_S$.

Open-circuit Voltage and Short-circuit Current

When the cell is operated at open circuit, $I = 0$ and the voltage across the output terminals is defined as the *open-circuit voltage*. Assuming the shunt resistance is high enough to neglect the final term of the characteristic equation, the open-circuit voltage V_{OC} is:

$$V_{OC} \approx \frac{nkT}{q} \ln\left(\frac{I_L}{I_0} + 1\right).$$

Similarly, when the cell is operated at short circuit, $V = 0$ and the current I through the terminals is defined as the *short-circuit current*. It can be shown that for a high-quality solar cell (low R_S and I_0, and high R_{SH}) the short-circuit current I_{SC} is:

$$I_{SC} \approx I_L.$$

It is not possible to extract any power from the device when operating at either open circuit or short circuit conditions.

Effect of Physical Size

The values of I_L, I_0, R_S, and R_{SH} are dependent upon the physical size of the solar cell. In comparing otherwise identical cells, a cell with twice the junction area of another will, in principle, have double the I_L and I_0 because it has twice the area where photocurrent is generated and across which diode current can flow. By the same argument, it will also have half the R_S of the series resistance related to vertical current flow; however, for large-area silicon solar cells, the scaling of the series resistance encountered by lateral current flow is not easily predictable since it will depend crucially on the grid design (it is not clear what "otherwise identical" means in this respect). Depending on the shunt type, the larger cell may also have half the R_{SH} because it has twice the area where shunts may occur; on the other hand, if shunts occur mainly at the perimeter, then R_{SH} will decrease according to the change in circumference, not area.

Since the changes in the currents are the dominating ones and are balancing each other, the open-circuit voltage is practically the same; V_{OC} starts to depend on the cell size only if R_{SH} becomes too low. To account for the dominance of the currents, the characteristic equation is frequently written in terms of current density, or current produced per unit cell area:

$$J = J_L - J_0\left\{\exp\left[\frac{q(V + Jr_S)}{nkT}\right] - 1\right\} - \frac{V + Jr_S}{r_{SH}}$$

where

- J = current density (ampere/cm²)

- J_L = photogenerated current density (ampere/cm²)

- J_o = reverse saturation current density (ampere/cm²)

- r_S = specific series resistance (Ω-cm²)

- r_{SH} = specific shunt resistance (Ω-cm²).

This formulation has several advantages. One is that since cell characteristics are referenced to a common cross-sectional area they may be compared for cells of different physical dimensions. While this is of limited benefit in a manufacturing setting, where all cells tend to be the same size, it is useful in research and in comparing cells between manufacturers. Another advantage is that the density equation naturally scales the parameter values to similar orders of magnitude, which can make numerical extraction of them simpler and more accurate even with naive solution methods.

There are practical limitations of this formulation. For instance, certain parasitic effects grow in importance as cell sizes shrink and can affect the extracted parameter values. Recombination and contamination of the junction tend to be greatest at the perimeter of the cell, so very small cells may exhibit higher values of J_o or lower values of R_{SH} than larger cells that are otherwise identical. In such cases, comparisons between cells must be made cautiously and with these effects in mind.

This approach should only be used for comparing solar cells with comparable layout. For instance, a comparison between primarily quadratical solar cells like typical crystalline silicon solar cells and narrow but long solar cells like typical thin film solar cells can lead to wrong assumptions caused by the different kinds of current paths and therefore the influence of, for instance, a distributed series resistance contribution to r_S. Macro-architecture of the solar cells could result in different surface areas being placed in any fixed volume - particularly for thin film solar cells and flexible solar cells which may allow for highly convoluted folded structures. If volume is the binding constraint, then efficiency density based on surface area may be of less relevance.

Cell Temperature

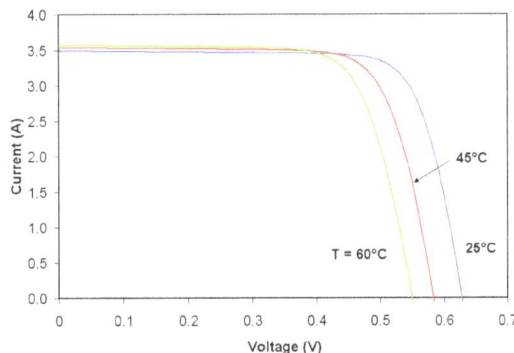

Effect of temperature on the current-voltage characteristics of a solar cell

Temperature affects the characteristic equation in two ways: directly, via T in the exponential term, and indirectly via its effect on I_o (strictly speaking, temperature affects all of the terms, but these two far more significantly than the others). While increasing T reduces the magnitude of the exponent in the characteristic equation, the value of I_o increases exponentially with T. The net effect is to reduce V_{OC} (the open-circuit voltage) linearly with increasing temperature. The magnitude of this reduction is inversely proportional to V_{OC}; that is, cells with higher values of V_{OC} suffer smaller reductions in voltage with increasing temperature. For most crystalline silicon solar cells the change in V_{OC} with temperature is about -0.50%/°C, though the rate for the highest-efficiency crystalline silicon cells is around -0.35%/°C. By way of comparison, the rate for amorphous silicon solar cells is -0.20%/°C to -0.30%/°C, depending on how the cell is made.

The amount of photogenerated current I_L increases slightly with increasing temperature because of an increase in the number of thermally generated carriers in the cell. This effect is slight, however: about 0.065%/°C for crystalline silicon cells and 0.09% for amorphous silicon cells.

The overall effect of temperature on cell efficiency can be computed using these factors in combination with the characteristic equation. However, since the change in voltage is much stronger than the change in current, the overall effect on efficiency tends to be similar to that on voltage. Most crystalline silicon solar cells decline in efficiency by 0.50%/°C and most amorphous cells decline by 0.15-0.25%/°C. The figure shows I-V curves that might typically be seen for a crystalline silicon solar cell at various temperatures.

Series Resistance

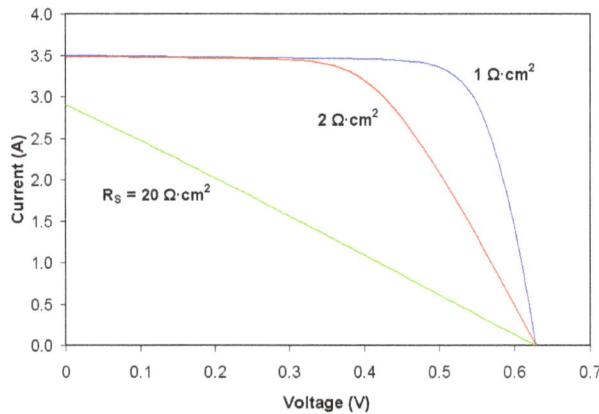

Effect of series resistance on the current-voltage characteristics of a solar cell.

As series resistance increases, the voltage drop between the junction voltage and the terminal voltage becomes greater for the same current. The result is that the current-controlled portion of the I-V curve begins to sag toward the origin, producing a significant decrease in the terminal voltage V and a slight reduction in I_{SC}, the short-circuit current. Very high values of R_S will also produce a significant reduction in I_{SC}; in these regimes, series resistance dominates and the behavior of the solar cell resembles that of a resistor. These effects are shown for crystalline silicon solar cells in the I-V curves displayed in the figure.

Losses caused by series resistance are in a first approximation given by $P_{loss}=V_{Rs}I=I^2R_S$ and increase quadratically with (photo-)current. Series resistance losses are therefore most important at high illumination intensities.

Shunt Resistance

As shunt resistance decreases, the current diverted through the shunt resistor increases for a given level of junction voltage. The result is that the voltage-controlled portion of the I-V curve begins to sag far from the origin, producing a significant decrease in the terminal current I and a slight reduction in V_{OC}. Very low values of R_{SH} will produce a significant reduction in V_{OC}. Much as in the case of a high series resistance, a badly shunted solar cell will take on operating characteristics similar to those of a resistor. These effects are shown for crystalline silicon solar cells in the I-V curves displayed in the figure.

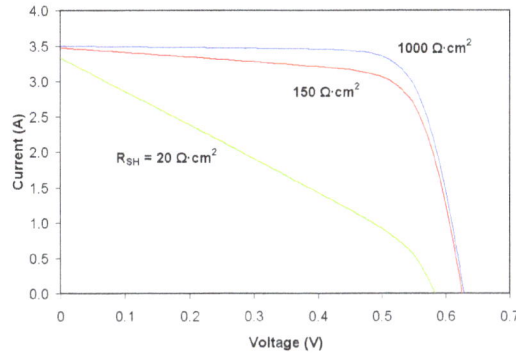

Effect of shunt resistance on the current–voltage characteristics of a solar cell.

Reverse Saturation Current

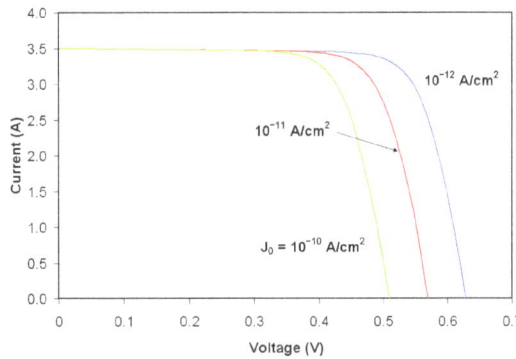

Effect of reverse saturation current on the current-voltage characteristics of a solar cell.

If one assumes infinite shunt resistance, the characteristic equation can be solved for V_{OC}:

$$V_{OC} = \frac{kT}{q} \ln\left(\frac{I_{SC}}{I_0} + 1\right).$$

Thus, an increase in I_o produces a reduction in V_{OC} proportional to the inverse of the logarithm of the increase. This explains mathematically the reason for the reduction in V_{OC} that accompanies increases in temperature described above. The effect of reverse saturation current on the I-V curve of a crystalline silicon solar cell are shown in the figure. Physically, reverse saturation current is a measure of the "leakage" of carriers across the p-n junction in reverse bias. This leakage is a result of carrier recombination in the neutral regions on either side of the junction.

Ideality Factor

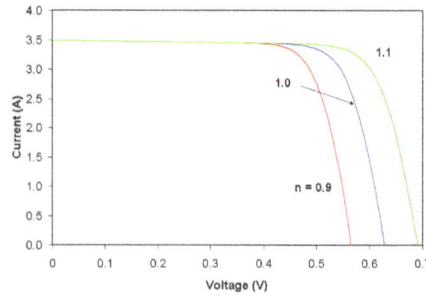

Effect of ideality factor on the current-voltage characteristics of a solar cell.

The ideality factor (also called the emissivity factor) is a fitting parameter that describes how closely the diode's behavior matches that predicted by theory, which assumes the p-n junction of the diode is an infinite plane and no recombination occurs within the space-charge region. A perfect match to theory is indicated when $n = 1$. When recombination in the space-charge region dominate other recombination, however, $n = 2$. The effect of changing ideality factor independently of all other parameters is shown for a crystalline silicon solar cell in the I-V curves displayed in the figure.

Most solar cells, which are quite large compared to conventional diodes, well approximate an infinite plane and will usually exhibit near-ideal behavior under Standard Test Condition ($n \approx 1$). Under certain operating conditions, however, device operation may be dominated by recombination in the space-charge region. This is characterized by a significant increase in I_o as well as an increase in ideality factor to $n \approx 2$. The latter tends to increase solar cell output voltage while the former acts to erode it. The net effect, therefore, is a combination of the increase in voltage shown for increasing n in the figure and the decrease in voltage shown for increasing I_o in the figure above. Typically, I_o is the more significant factor and the result is a reduction in voltage.

Sometimes, the ideality factor is observed to be greater than 2, which is generally attributed to the presence of Schottky diode or heterojunction in the solar cell. The presence of a heterojunction offset reduces the collection efficiency of the solar cell and may contribute to low fill-factor.

Solar Cell Efficiency

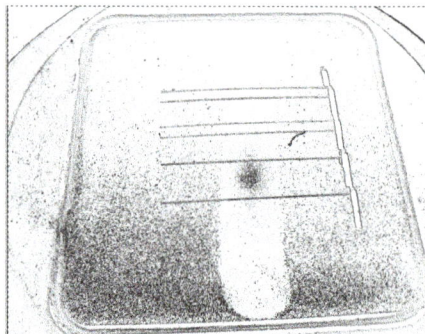

Dust often accumulates on the glass of solar modules - seen here as black dots
- which reduces the amount of light admitted to the solar cells.

Solar cell efficiency refers to the portion of energy in the form of sunlight that can be converted via photovoltaics into electricity.

The efficiency of the solar cells used in a photovoltaic system, in combination with latitude and climate, determines the annual energy output of the system. For example, a solar panel with 20% efficiency and an area of 1 m² will produce 200 W at Standard Test Conditions, but it can produce more when the sun is high in the sky and will produce less in cloudy conditions and when the sun is low in the sky. In central Colorado, which receives annual insolation of 5.5 kWh/m²/day, such a panel can be expected to produce 440 kWh of energy per year. However, in Michigan, which receives only 3.8 kWh/m²/day, annual energy yield will drop to 280 kWh for the same panel. At more northerly European latitudes, yields are significantly lower: 175 kWh annual energy yield in southern England.

Several factors affect a cell's conversion efficiency value, including its reflectance efficiency, thermodynamic efficiency, charge carrier separation efficiency, and conduction efficiency values. Because these parameters can be difficult to measure directly, other parameters are measured instead, including quantum efficiency, V_{OC} ratio, and fill factor. Reflectance losses are accounted for by the quantum efficiency value, as they affect "external quantum efficiency." Recombination losses are accounted for by the quantum efficiency, V_{OC} ratio, and fill factor values. Resistive losses are predominantly accounted for by the fill factor value, but also contribute to the quantum efficiency and V_{OC} ratio values.

As of December 2014, the world record for solar cell efficiency at 46% was achieved by using multi-junction concentrator solar cells, developed from collaboration efforts of Soitec, CEA-Leti, France together with Fraunhofer ISE, Germany.

Factors Affecting Energy Conversion Efficiency

The factors affecting energy conversion efficiency were expounded in a landmark paper by William Shockley and Hans Queisser in 1961.

Thermodynamic Efficiency Limit and Infinite-stack Limit

If one has a source of heat at temperature T_s and cooler heat sink at temperature T_c, the maximum theoretically possible value for the ratio of work (or electric power) obtained to heat supplied is $1-T_c/T_s$, given by a Carnot heat engine. If we take 6000 K for the temperature of the sun and 300 K for ambient conditions on earth, this comes to 95%. In 1981, Alexis de Vos and Herman Pauwels showed that this is achievable with a stack of an infinite number of cells with band gaps ranging from infinity (the first cells encountered by the incoming photons) to zero, with a voltage in each cell very close to the open-circuit voltage, equal to 95% of the band gap of that cell, and with 6000 K blackbody radiation coming from all directions. However, the 95% efficiency thereby achieved means that the electric power is 95% of the *net* amount of light absorbed – the stack *emits* radiation as though it were almost at 6000 K, and this radiation has to be subtracted from the incoming radiation when calculating the amount of heat being transferred and the efficiency. They also considered the more relevant problem of maximizing the power output for a stack being illuminated from all directions by 6000 K blackbody radiation. In this case, the voltages must be lowered to less than 95% of the band gap (the percentage is not constant over all the cells). They found that the maximum power was 86.8% of the amount of *in-coming* radiation. When the in-coming radiation comes only from an area of the sky the size of the sun, the efficiency limit drops to 68.7%.

Ultimate Efficiency

Normal photovoltaic systems however have only one p-n junction and are therefore subject to a lower efficiency limit, called the "ultimate efficiency" by Shockley and Queisser. Photons with an energy below the band gap of the absorber material cannot generate an electron-hole pair, so their energy is not converted to useful output, and only generates heat if absorbed. For photons with an energy above the band gap energy, only a fraction of the energy above the band gap can be converted to useful output. When a photon of greater energy is absorbed, the excess energy above the band gap is converted to kinetic energy of the carrier combination. The excess kinetic energy is converted to heat through phonon interactions as the kinetic energy of the carriers slows to equilibrium velocity. Traditional single-junction cells have a maximum theoretical efficiency of 33.16%.

Solar cells with multiple band gap absorber materials improve efficiency by dividing the solar spectrum into smaller bins where the thermodynamic efficiency limit is higher for each bin.

Quantum Efficiency

As described above, when a photon is absorbed by a solar cell it can produce an electron-hole pair. One of the carriers may reach the p-n junction and contribute to the current produced by the solar cell; such a carrier is said to be *collected*. Or, the carriers recombine with no net contribution to cell current.

Quantum efficiency refers to the percentage of photons that are converted to electric current (i.e., collected carriers) when the cell is operated under short circuit conditions. The "external" quantum efficiency of a silicon solar cell includes the effect of optical losses such as transmission and reflection.

In particular, some measures can be taken to reduce these losses. The reflection losses, which can account for up to 10% of the total incident energy, can be dramatically decreased using a technique called texturization, a light trapping method that modifies the average light path.

Quantum efficiency is most usefully expressed as a *spectral* measurement (that is, as a function of photon wavelength or energy). Since some wavelengths are absorbed more effectively than others, spectral measurements of quantum efficiency can yield valuable information about the quality of the semiconductor bulk and surfaces. Quantum efficiency alone is not the same as overall energy conversion efficiency, as it does not convey information about the fraction of power that is converted by the solar cell.

Maximum Power Point

A solar cell may operate over a wide range of voltages (V) and currents (I). By increasing the resistive load on an irradiated cell continuously from zero (a *short circuit*) to a very high value (an *open circuit*) one can determine the maximum power point, the point that maximizes V×I; that is, the load for which the cell can deliver maximum electrical power at that level of irradiation. (The output power is zero in both the short circuit and open circuit extremes).

A high quality, monocrystalline silicon solar cell, at 25 °C cell temperature, may produce 0.60 V open-circuit (V_{OC}). The cell temperature in full sunlight, even with 25 °C air temperature, will probably be close to 45 °C, reducing the open-circuit voltage to 0.55 V per cell. The voltage drops modestly, with this type of cell, until the short-circuit current is approached (I_{SC}). Maximum power (with 45 °C cell temperature) is typically produced with 75% to 80% of the open-circuit voltage (0.43 V in this case) and 90% of the short-circuit current. This output can be up to 70% of the $V_{OC} \times I_{SC}$ product. The short-circuit current (I_{SC}) from a cell is nearly proportional to the illumination, while the open-circuit voltage (V_{OC}) may drop only 10% with an 80% drop in illumination. Lower-quality cells have a more rapid drop in voltage with increasing current and could produce only $1/2\ V_{OC}$ at $1/2\ I_{SC}$. The usable power output could thus drop from 70% of the $V_{OC} \times I_{SC}$ product to 50% or even as little as 25%. Vendors who rate their solar cell "power" only as $V_{OC} \times I_{SC}$, without giving load curves, can be seriously distorting their actual performance.

The maximum power point of a photovoltaic varies with incident illumination. For example, accumulation of dust on photovoltaic panels reduces the maximum power point. For systems large enough to justify the extra expense, a maximum power point tracker tracks the instantaneous power by continually measuring the voltage and current (and hence, power transfer), and uses this information to dynamically adjust the load so the maximum power is *always* transferred, regardless of the variation in lighting.

Fill Factor

Another defining term in the overall behavior of a solar cell is the *fill factor* (*FF*). This is the available *power* at the *maximum power point* (P_m) divided by the *open circuit voltage* (V_{OC}) and the *short circuit current* (I_{SC}):

$$FF = \frac{P_m}{V_{OC} \times I_{SC}} = \frac{\eta \times A_c \times G}{V_{OC} \times I_{SC}}.$$

The fill factor is directly affected by the values of the cell's series, shunt resistances and diodes losses. Increasing the shunt resistance (R_{sh}) and decreasing the series resistance (R_s) lead to a higher fill factor, thus resulting in greater efficiency, and bringing the cell's output power closer to its theoretical maximum.

The Fill factor has a value around 80% for a normal silicon PV cell.

Comparison

Energy conversion efficiency is measured by dividing the electrical output by the incident light power. Factors influencing output include spectral distribution, spatial distribution of power, temperature, and resistive load. IEC standard 61215 is used to compare the performance of cells and is designed around standard (terrestrial, temperate) temperature and conditions (STC): irradiance of 1 kW/m², a spectral distribution close to solar radiation through AM (airmass) of 1.5 and a cell temperature 25 °C. The resistive load is varied until the peak or maximum power point (MPP) is achieved. The power at this point is recorded as Watt-peak (Wp). The same standard is used for measuring the power and efficiency of PV modules.

Air mass affects output. In space, where there is no atmosphere, the spectrum of the sun is relatively unfiltered. However, on earth, air filters the incoming light, changing the solar spectrum. The filtering effect ranges from Air Mass 0 (AM0) in space, to approximately Air Mass 1.5 on Earth. Multiplying the spectral differences by the quantum efficiency of the solar cell in question yields the efficiency. Terrestrial efficiencies typically are greater than space efficiencies. For example, a silicon solar cell in space might have an efficiency of 14% at AM0, but 16% on earth at AM 1.5. Note, however, that incident photons in space carry considerably more energy, so the solar cell might produce considerably more power in space, despite the lower efficiency as indicated by reduced percentage of the total incident energy captured.

Solar cell efficiencies vary from 6% for amorphous silicon-based solar cells to 44.0% with multiple-junction production cells and 44.4% with multiple dies assembled into a hybrid package. Solar cell energy conversion efficiencies for commercially available *multicrystalline Si* solar cells are around 14-19%. The highest efficiency cells have not always been the most economical — for example a 30% efficient multijunction cell based on exotic materials such as gallium arsenide or indium selenide produced at low volume might well cost one hundred times as much as an 8% efficient amorphous silicon cell in mass production, while delivering only about four times the output.

However, there is a way to "boost" solar power. By increasing the light intensity, typically photo-generated carriers are increased, increasing efficiency by up to 15%. These so-called "concentrator systems" have only begun to become cost-competitive as a result of the development of high efficiency GaAs cells. The increase in intensity is typically accomplished by using concentrating optics. A typical concentrator system may use a light intensity 6-400 times the sun, and increase the efficiency of a one sun GaAs cell from 31% at AM 1.5 to 35%.

A common method used to express economic costs is to calculate a price per delivered kilowatt-hour (kWh). The solar cell efficiency in combination with the available irradiation has a major influence on the costs, but generally speaking the overall system efficiency is important. Commercially available solar cells (as of 2006) reached system efficiencies between 5 and 19%.

Undoped crystalline silicon devices are approaching the theoretical limiting efficiency of 29.4% In 2014, efficiency of 25.6% was achieved in crystalline cells that place both positive and negative contacts on the back of the cell and that cover the wafer's front and back with thin films of silicon.

Energy Payback

The energy payback time is defined as the recovery time required for generating the energy spent for manufacturing a modern photovoltaic module. In 2008 it was estimated to be from 1 to 4 years depending on the module type and location. With a typical lifetime of 20 to 30 years, this means that, modern solar cells would be net energy producers, i.e. they would generate more energy over their lifetime than the energy expended in producing them. Generally, thin-film technologies—despite having comparatively low conversion efficiencies—achieve significantly shorter energy payback times than conventional systems (often < 1 year).

A study published in 2013 which the existing literature found that energy payback time was between 0.75 and 3.5 years with thin film cells being at the lower end and multi-si-cells having a payback time of 1.5-2.6 years. A 2015 review assessed the energy payback time and EROI of solar photovoltaics. In this meta study, which uses an insolation of 1700 kWh/m²/year and a system lifetime of 30 years, mean harmonized EROIs between 8.7 and 34.2 were found. Mean harmonized energy payback time varied from 1.0 to 4.1 years. Crystalline silicon devices achieve on average an energy payback period of 2 years.

Like any other technology, solar cell manufacture is dependent on and presupposes the existence of a complex global industrial manufacturing system. This comprises not only the fabrication systems typically accounted for in estimates of manufacturing energy, but the contingent mining, refining and global transportation systems, as well as other energy intensive critical support systems including finance, information, and security systems. The uncertainty of that energy component confers uncertainty on any estimate of payback times derived from that estimate, considered by some to be significant.

Technical Methods of Improving Efficiency

Promoting Light Scattering in the Visible Spectrum

By lining the light receiving surface of the cell with nano-sized metallic studs, the efficiency of the cell can be substantially increased, as the light reflects off these studs at an oblique angle to the cell, increasing the length of the path the light takes through the cell, thereby increasing the number of photons absorbed by the cell, and so also the amount of current generated.

The main materials used for the nano-studs are silver, gold, and aluminium, to name a few. However, gold and silver are not very efficient, as they absorb much of the light in the visible spectrum, which contains most of the energy present in sunlight, reducing the amount of light reaching the cell. Aluminium, on the other hand, absorbs only ultraviolet radiation, and reflects both visible and infra-red light, so energy loss is minimized on that front. Aluminium is therefore capable of increasing the efficiency of the cell by up to 22% (in lab conditions).

Radiative Cooling

An increase in solar cell temperature of around 1 °C leads to a decrease in efficiency of about 0.45%.To prevent decreased efficiency due to heating, a visibly transparent silica crystal layer can be applied to a solar panel. The silica layer acts as a thermal black body which emits heat as infra-red radiation into space cooling the cell by up to 13 °C.

Solar Panels on Spacecraft

Spacecraft operating in the inner Solar System usually rely on the use of photovoltaic solar panels to derive electricity from sunlight. In the outer solar system, where the sunlight is too weak to produce sufficient power, radioisotope thermoelectric generators (RTGs) are used as a power source.

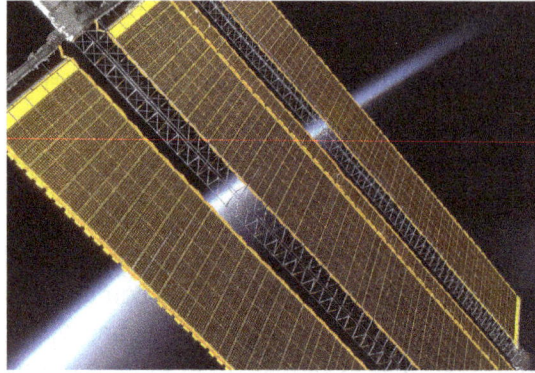

A solar panel array of the International Space Station (Expedition 17 crew, August 2008).

History

The first spacecraft to use solar panels was the Vanguard 1 satellite, Launched by the US in 1958. This was largely because of the influence of Dr. Hans Ziegler, who can be regarded as the father of spacecraft SOLAR power.

Uses

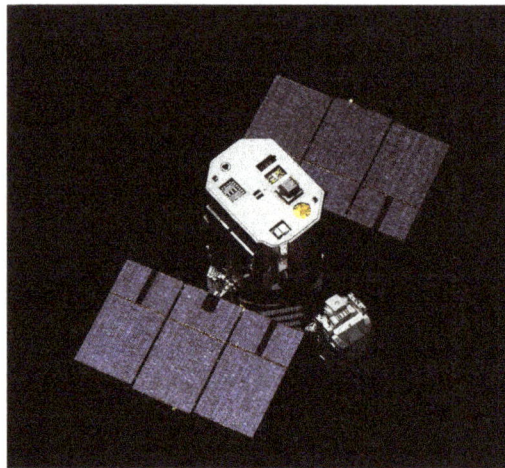

The SOLAR panels on the SMM satellite provided electrical power. Here it is being captured by an astronaut in a mobile space-suit that runs on chemical battery power.

Solar panels on spacecraft supply power for two main uses:

- power to run the sensors, active heating, cooling and telemetry.

- power for spacecraft propulsion – electric propulsion, sometimes called solar-electric propulsion.

For both uses, a key figure of merit of the solar panels is the specific power (watts generated divided by solar array mass), which indicates on a relative basis how much power one array will generate for a given launch mass relative to another. Another key metric is stowed packing efficiency (deployed watts produced divided by stowed volume), which indicates how easily the array will fit into a launch vehicle. Yet another key metric is cost (dollars per watt).

To increase the specific power, typical solar panels on spacecraft use close-packed solar cell rectangles that cover nearly 100% of the sun-visible area of the solar panels, rather than the solar wafer circles which, even though close-packed, cover about 90% of the sun-visible area of typical solar panels on earth. However, some solar panels on spacecraft have solar cells that cover only 30% of the sun-visible area.

Implementation

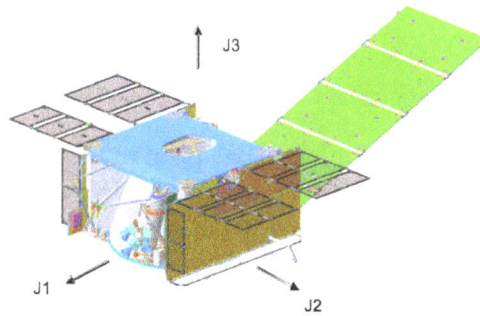

Diagram of the Spacecraft Bus on the planned James Webb Space Telescope, which is powered by solar panels (coloured green in this 3/4 view). Note that shorter light purple extensions are radiator shades not solar panels.

Solar panels need to have a lot of surface area that can be pointed towards the Sun as the spacecraft moves. More exposed surface area means more electricity can be converted from light energy from the Sun. Since spacecraft have to be small, this limits the amount of power that can be produced.

All electrical circuits generate waste heat; in addition, solar arrays act as optical and thermal as well as electrical collectors. Heat must be radiated from their surfaces. High-power spacecraft may have solar arrays that compete with the active payload itself for thermal dissipation. The innermost panel of arrays may be "blank" to reduce the overlap of views to space. Such spacecraft include the higher-power communications satellites (e.g., later-generation TDRS) and Venus Express, not high-powered but closer to the Sun.

Spacecraft are built so that the solar panels can be pivoted as the spacecraft moves. Thus, they can always stay in the direct path of the light rays no matter how the spacecraft is pointed. Spacecraft are usually designed with solar panels that can always be pointed at the Sun, even as the rest of the body of the spacecraft moves around, much as a tank turret can be aimed independently of where the tank is going. A tracking mechanism is often incorporated into the solar arrays to keep the array pointed towards the sun.

Sometimes, satellite operators purposefully orient the solar panels to "off point," or out of direct alignment from the Sun. This happens if the batteries are completely charged and the amount of electricity needed is lower than the amount of electricity made; off-pointing is also sometimes used on the International Space Station for orbital drag reduction.

Ionizing Radiation Issues and Mitigation

Space contains varying levels of ionizing radiation, that which includes flares and other solar events. Some satellites orbit within the protective zone of the magnetosphere, while others do not.

Types of Solar Cells Typically Used

Gallium arsenide-based solar cells are typically favored over crystalline silicon in industry because they have a higher efficiency and degrade more slowly than silicon in the radiation present in space. The most efficient solar cells currently in production are multi-junction photovoltaic cells. These use a combination of several layers of gallium arsenide, indium gallium phosphide, and germanium to capture more energy from the solar spectrum. Leading edge multi-junction cells are capable of exceeding 38.8% under non-concentrated AM1.5G illumination and 46% using concentrated AM1.5G illumination.

Spacecraft that have used Solar Power

To date, solar power, other than for propulsion, has been practical for spacecraft operating no farther from the Sun than the orbit of Jupiter. For example, Juno, Magellan, Mars Global Surveyor, and Mars Observer used solar power as does the Earth-orbiting, Hubble Space Telescope. The Rosetta space probe, launched 2 March 2004, used its 64 square metres (690 sq ft) of solar panels as far as the orbit of Jupiter (5.25 AU); previously the furthest use was the Stardust spacecraft at 2 AU. Solar power for propulsion was also used on the European lunar mission SMART-1 with a Hall effect thruster.

The *Juno* mission, launched in 2011, is the first mission to Jupiter (arrived at Jupiter on July 4, 2016) to use solar panels instead of the traditional RTGs that are used by previous outer solar system missions, making it the furthest spacecraft to use solar panels to date. It has 72 square metres (780 sq ft) of panels.

Another spacecraft of interest is Dawn which went into orbit around 4 Vesta in 2011. It used ion thrusters to get to Ceres.

The potential for solar powered spacecraft beyond Jupiter has been studied.

Power Available

In 2005 Rigid-Panel Stretched Lens Arrays were producing 7 kW per wing. Solar arrays producing 300 W/kg and 300 W/m^2 from the sun's 1366 W/m^2 power near the Earth are available.

Future Uses

For future missions, it is desirable to reduce solar array mass, and to increase the power generated per unit area. This will reduce overall spacecraft mass, and may make the operation of solar-powered spacecraft feasible at larger distances from the sun. Solar array mass could be reduced with thin-film photovoltaic cells, flexible blanket substrates, and composite support structures. Solar array efficiency could be improved by using new photovoltaic cell materials and solar concentrators that intensify the incident sunlight. Photovoltaic concentrator solar arrays for primary spacecraft power are devices which intensify the sunlight on the photovoltaics. This design uses a flat lens, called a Fresnel lens, which takes a large area of sunlight and concentrates it onto a smaller spot. The same principle is used to start fires with a magnifying glass on a sunny day.

Solar concentrators put one of these lenses over every solar cell. This focuses light from the large concentrator area down to the smaller cell area. This allows the quantity of expensive solar cells to be reduced by the amount of concentration. Concentrators work best when there is a single source of light and the concentrator can be pointed right at it. This is ideal in space, where the Sun is a single light source. Solar cells are the most expensive part of solar arrays, and arrays are often a very expensive part of the spacecraft. This technology may allow costs to be cut significantly due to the utilization of less material.

References

- Kim, H. C., Fthenakis, V., Choi, J. K. & Turney, D. E. (2012). "Life cycle greenhouse gas emissions of thin-film photovoltaic electricity generation". Journal of Industrial Ecology. 16: S110–S121. doi:10.1111/j.1530-9290.2011.00423.x

- Liebreich, Michael (29 January 2014). "A YEAR OF CRACKING ICE: 10 PREDICTIONS FOR 2014". Bloomberg New Energy Finance. Retrieved 24 April 2014

- Gong, J., Darling, S. B. & You, F. (2015). "Perovskite photovoltaics: life-cycle assessment of energy and environmental impacts". Energy & Environmental Science. 8 (7): 1953–1968. doi:10.1039/C5EE00615E

- Palz, Wolfgang (2013). Solar Power for the World: What You Wanted to Know about Photovoltaics. CRC Press. pp. 131–. ISBN 978-981-4411-87-5

- "2014 Outlook: Let the Second Gold Rush Begin" (PDF). Deutsche Bank Markets Research. 6 January 2014. Archived from the original on 22 November 2014. Retrieved 22 November 2014

- Brown, G. F. & Wu, J. (2009). "Third generation photovoltaics". Laser & Photonics Reviews. 3 (4): 394–405. doi:10.1002/lpor.200810039

- Heidari, N., Gwamuri, J., Townsend, T.,Pearce, J.M. (2015). open access Impact of Snow and Ground Interference on Photovoltaic Electric System Performance. IEEE Journal of Photovoltaics 5(6),1680-1685, (2015)

- Luque, Antonio & Hegedus, Steven (2003). Handbook of Photovoltaic Science and Engineering. John Wiley and Sons. ISBN 0-471-49196-9

- Cole, IR, Betts, TR, Gottschalg, R (2012), "Solar profiles and spectral modeling for CPV simulations", IEEE Journal of Photovoltaics, 2 (1): 62–67, ISSN 2156-3381, doi:10.1109/JPHOTOV.2011.2177445

- "UD-led team sets solar cell record, joins DuPont on $100 million project". UDaily. University ofDelaware. 24 July 2007. Retrieved 24 July 2007

- Espinosa, Nieves; García-Valverde, Rafael; Urbina, Antonio; Krebs, Frederik C. (2011). "A life cycle analysis of polymer solar cell modules prepared using roll-to-roll methods under ambient conditions". Solar Energy Materials and Solar Cells. 95 (5): 1293–1302. doi:10.1016/j.solmat.2010.08.020

Energy Conversion: An Integrated Study

Energy conversion is the act of changing energy. The change that occurs converts energy from one form to another. This energy can be harvested to ultimately power devices. The aspects elucidated in this chapter are of vital importance, and provide a better understanding of energy conversion.

Energy Conversion

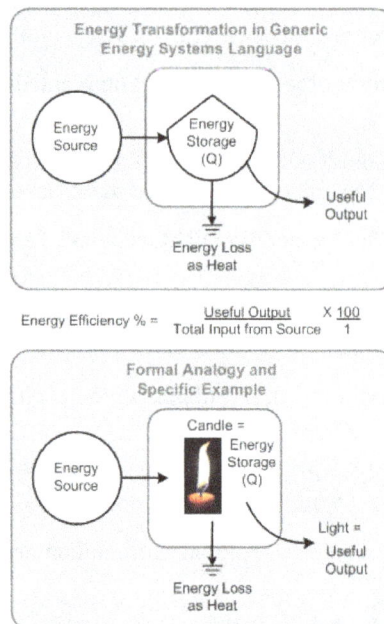

Energy Transformation in Energy Systems Language.

Energy transformation, also termed energy conversion, is the process of changing energy from one of its forms into another. In physics, energy is a quantity that provides the capacity to perform many actions—some as simple as lifting or warming an object. In addition to being convertible, energy is transferable to a different location or object, but it cannot be created or destroyed.

Energy in many of its forms may be used in natural processes, or to provide some service to society such as heating, refrigeration, lighting, or performing mechanical work to operate machines. For example, in order to heat your home, your furnace can burn fuel, whose chemical potential energy is thus converted into thermal energy, which is then transferred to your home's air in order to raise its temperature.

In another example, an internal combustion engine burns gasoline to cause pressure that pushes the pistons, thus performing work in order to accelerate your vehicle, ultimately converting the fuel's chemical energy to your vehicle's additional kinetic energy corresponding to its increase in speed.

Entropy and Limitations in Conversion of Thermal Energy to Other Types

Conversions to thermal energy (thus raising the temperature) from other forms of energy, may occur with essentially 100% efficiency (many types of friction do this). Conversion among non-thermal forms of energy may occur with fairly high efficiency, though there is always some energy dissipated thermally due to friction and similar processes. Sometimes the efficiency is close to 100%, such as when potential energy is converted to kinetic energy as an object falls in vacuum, or when an object orbits nearer or farther from another object, in space.

Though, conversion of thermal energy to other forms, thus reducing the temperature of a system, has strict limitations, often keeping its efficiency much less than 100% (even when energy is not allowed to escape from the system). This is because thermal energy has already been partly spread out among many available states of a collection of microscopic particles constituting the system, which can have enormous numbers of possible combinations of momentum and position (these combinations are said to form a phase space). In such circumstances, a measure called entropy, or evening-out of energy distributions, dictates that future states of an isolated system must be of at least equal evenness in energy distribution. In other words, there is no way to *concentrate* energy without spreading out energy somewhere else.

Thermal energy in equilibrium at a given temperature already represents the maximal evening-out of energy between all possible states. Such energy is sometimes considered "degraded energy," because it is not entirely convertible a "useful" form, i.e. one that can do more than just affect temperature. The second law of thermodynamics is a way of stating that, for this reason, thermal energy in a system may be converted to other kinds of energy with efficiencies approaching 100%, only if the entropy (even-ness or disorder) of the universe is increased by other means, to compensate for the decrease in entropy associated with the disappearance of the thermal energy and its entropy content. Otherwise, only a part of thermal energy may be converted to other kinds of energy (and thus, useful work), since the remainder of the heat must be reserved to be transferred to a thermal reservoir at a lower temperature, in such a way that the increase in Entropy for this process more than compensates for the entropy decrease associated with transformation of the rest of the heat into other types of energy.

Transformation of Kinetic Energy of Charged Particles to Electric Energy

In order to make the energy transformation more efficient, it is desirable to avoid the thermal conversion. For example, the efficiency of nuclear energy reactors, where kinetic energy of nuclei is first converted to thermal energy and then to electric energy, lies around 35%. By direct conversion of kinetic energy to electric, i.e. by eliminating the thermal energy transformation, the efficiency of energy transformation process can be dramatically improved.

History of Energy Transformation from the Early Universe

Energy transformations in the universe over time are (generally) characterized by various kinds of energy which has been available since the Big Bang, later being "released" (that is, transformed to more active types of energy such as kinetic or radiant energy), when a triggering mechanism is available to do it.

Release of energy from gravitational potential: A direct transformation of energy occurs when hydrogen produced in the big bang collects into structures such as planets, in a process during which part of the gravitational potential is to be converted directly into heat. In Jupiter, Saturn, and Neptune, for example, such heat from continued collapse of the planets' large gas atmospheres continues to drive most of the planets' weather systems, with atmospheric bands, winds, and powerful storms which are only partly powered by sunlight, however, on Uranus, little of this process occurs.

On Earth a significant portion of heat output from interior of the planet, estimated at a third to half of the total, is caused by slow collapse of planetary materials to a smaller size, with output of gravitationally driven heat.

Release of energy from radioactive potential: Familiar examples of other such processes transforming energy from the Big Bang include nuclear decay, in which energy is released which was originally "stored" in heavy isotopes, such as uranium and thorium. This energy was stored at the time of these elements' nucleosynthesis, a process which ultimately uses the gravitational potential energy released from the gravitational collapse of type IIa supernovae, to store energy in the creation of these heavy elements before they were incorporated into the solar system and the Earth. Such energy locked into uranium is triggered for sudden-release in nuclear fission bombs, and similar stored energies in atomic nuclei are released spontaneously, during most types of radioactive decay. In such processes, heat from decay of these atoms of radioisotope in the core of the Earth is transformed immediately to heat. This heat in turn may lift mountains, via plate tectonics and orogenesis. This slow lifting of terrain thus represents a kind of gravitational potential energy storage of the heat energy. The stored potential energy may be released to active kinetic energy in landslides, after a triggering event. Earthquakes also release stored elastic potential energy in rocks, a kind of mechanical potential energy which has been produced ultimately from the same radioactive heat sources.

Thus, according to present understanding, familiar events such as landslides and earthquakes release energy which has been stored as potential energy in the Earth's gravitational field, or elastic strain (mechanical potential energy) in rocks. Prior to this, the energy represented by these events had been stored in heavy atoms (or in the gravitational potential of the Earth). The energy stored in heat atoms had been stored as potential ever since the time that gravitational potentials transforming energy in the collapse of long-destroyed stars (supernovae) created these atoms, and in doing so, stored the energy within.

Release of energy from hydrogen fusion potential: In other similar chain of transformations beginning at the dawn of the universe, nuclear fusion of hydrogen in the Sun releases another store of potential energy which was created at the time of the Big Bang. At that time, according to theory, space expanded and the universe cooled too rapidly for hydrogen to completely fuse into heavier elements. This resulted in hydrogen representing a store of potential energy which can be released by nuclear fusion. Such a fusion process is triggered by heat and pressure generated from gravitational collapse of hydrogen clouds when they produce stars, and some of the fusion energy is then transformed into sunlight. Such sunlight may again be stored as gravitational potential energy after it strikes the Earth, as (for example) snow-avalanches, or when water evaporates from oceans and is deposited high above sea level (where, after being released at a hydroelectric dam, it can be used to drive turbine/generators to produce electricity). Sunlight also drives many weather phenomena on Earth. An example of a solar-mediated weather event is a hurricane, which occurs when large unstable areas of warm ocean, heated over months, give up some of their thermal energy suddenly to power a few days of violent air movement. Sunlight is also captured by green plants as *chemical potential energy*, when carbon dioxide and water are converted into a combustible combination of carbohydrates, lipids, and oxygen. Release of this energy as heat and light may be triggered suddenly by a spark, in a forest fire; or it may be available more slowly for animal or human metabolism, when these molecules are ingested, and catabolism is triggered by enzyme action.

Through all of these transformation chains, potential energy stored at the time of the Big Bang is later released by intermediate events, sometimes being stored in a number of ways over time between releases, as more active energy. In all these events, one kind of energy is converted to other types of energy, including heat.

Examples of Sets of Energy Conversions in Machines

For instance, a coal-fired power plant involves these energy transformations:

1. Chemical energy in the coal converted to thermal energy in the exhaust gases of combustion.

2. Thermal energy of the exhaust gases converted into thermal energy of steam through the heat exchanger.

3. Thermal energy of steam converted to mechanical energy in the turbine.

4. Mechanical energy of the turbine converted to electrical energy by the generator, which is the ultimate output

In such a system, the first and fourth step are highly efficient, but the second and third steps are less efficient. The most efficient gas-fired electrical power stations can achieve 50% conversion efficiency. Oil- and coal-fired stations achieve less.

In a conventional automobile, these energy transformations are involved:

1. Chemical energy in the fuel converted to kinetic energy of expanding gas via combustion

2. Kinetic energy of expanding gas converted to linear piston movement

3. Linear piston movement converted to rotary crankshaft movement

4. Rotary crankshaft movement passed into transmission assembly

5. Rotary movement passed out of transmission assembly

6. Rotary movement passed through differential

7. Rotary movement passed out of differential to drive wheels

8. Rotary movement of drive wheels converted to linear motion of the vehicle.

Other Energy Conversions

There are many different machines and transducers that convert one energy form into another. A short list of examples follows:

- Thermoelectric (Heat → Electric energy)
- Geothermal power (Heat→ Electric energy)
- Heat engines, such as the internal combustion engine used in cars, or the steam engine (Heat → Mechanical energy)
- Ocean thermal power (Heat → Electric energy)
- Hydroelectric dams (Gravitational potential energy → Electric energy)
- Electric generator (Kinetic energy or Mechanical work → Electric energy)
- Fuel cells (Chemical energy → Electric energy)
- Battery (electricity) (Chemical energy → Electric energy)
- Fire (Chemical energy → Heat and Light)
- Electric lamp (Electric energy → Heat and Light)
- Microphone (Sound → Electric energy)
- Wave power (Mechanical energy → Electric energy)
- Windmills (Wind energy → Electric energy or Mechanical energy)
- Piezoelectrics (Strain → Electric energy)
- Friction (Kinetic energy → Heat)
- Electric heater (Electric energy → Heat)

Energy Conversion Efficiency

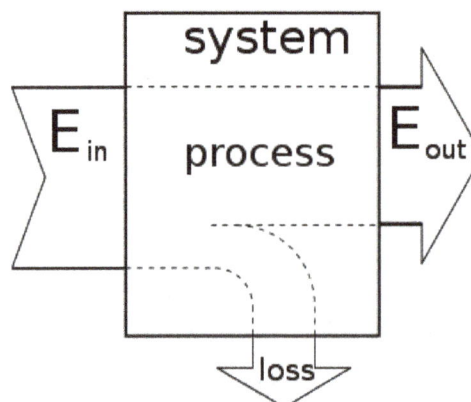

Output energy is always lower than input energy

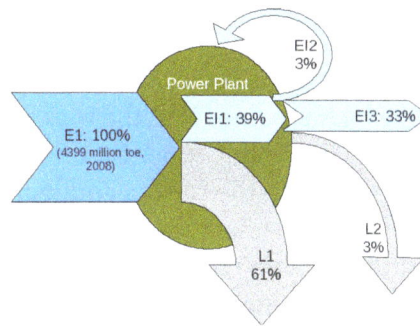

Efficiency of Power Plant, World total 2008

Energy conversion efficiency (η) is the ratio between the useful output of an energy conversion machine and the input, in energy terms. The input, as well as the useful output may be electric power, mechanical work, light (radiation), or heat.

Overview

Energy conversion efficiency is not defined uniquely, but instead depends on the usefulness of the output. All or part of the heat produced from burning a fuel may become rejected waste heat if, for example, work is the desired output from a thermodynamic cycle. Energy converter is an example of an energy transformation. For example a light bulb falls into the categories energy converter. $\eta = \dfrac{P_{out}}{P_{in}}$ Even though the definition includes the notion of usefulness, efficiency is considered a technical or physical term. Goal or mission oriented terms include effectiveness and efficacy.

Generally, energy conversion efficiency is a dimensionless number between 0 and 1.0, or 0% to 100%. Efficiencies may not exceed 100%, e.g., for a perpetual motion machine. However, other effectiveness measures that can exceed 1.0 are used for heat pumps and other devices that move heat rather than convert it.

When talking about the efficiency of heat engines and power stations the convention should be stated, i.e., HHV (a.k.a. Gross Heating Value, etc.) or LCV (a.k.a. Net Heating value), and whether gross output (at the generator terminals) or net output (at the power station fence) are being considered. The two are separate but both must be stated. Failure to do so causes endless confusion.

Related, more specific terms include

- Electrical efficiency, useful power output per electrical power consumed;

- Mechanical efficiency, where one form of mechanical energy (e.g. potential energy of water) is converted to mechanical energy (work);

- Thermal efficiency or Fuel efficiency, useful heat and/or work output per input energy such as the fuel consumed;

- 'Total efficiency', e.g., for cogeneration, useful electric power and heat output per fuel energy consumed. Same as the thermal efficiency.

- Luminous efficiency, that portion of the emitted electromagnetic radiation is usable for human vision.

Fuel Heating Values and Efficiency

In Europe the usable energy content of fuel is typically calculated using the lower heating value (LHV) of that fuel, the definition of which assumes that the water vapor produced during fuel combustion (oxidation), remains gaseous, and is not condensed to liquid water so the latent heat of vaporization of that water is not usable. Using the LHV, a condensing boiler can achieve a "heating efficiency" in excess of 100% (this does not violate the first law of thermodynamics as long as the LHV convention is understood, but does cause confusion). This is because the apparatus recovers part of the heat of vaporization, which is not included in the definition of the lower heating value of fuel. In the U.S. and elsewhere, the higher heating value (HHV) is used, which includes the latent heat for condensing the water vapor, and thus the thermodynamic maximum of 100% efficiency cannot be exceeded with HHV's use.

Example of Energy Conversion Efficiency

Conversion process	Conversion type	Energy efficiency
Electricity generation		
Gas turbine	Chemical to electrical	up to 40%
Gas turbine plus steam turbine (combined cycle)	Chemical/thermal to electrical	up to 60%
Water turbine	Gravitational to electrical	up to 90% (practically achieved)
Wind turbine	Kinetic to electrical	up to 59% (theoretical limit)
Solar cell	Radiative to electrical	6–40% (technology-dependent, 15-20% most often, 85–90% theoretical limit)
Fuel cell	Chemical to electrical	up to 85%
World Electricity generation 2008	Gross output 39%	Net output 33%
Electricity storage		
Lithium-ion battery	Chemical to electrical/reversible	80–90%
Nickel-metal hydride battery	Chemical to electrical/reversible	66%
Lead-acid battery	Chemical to electrical/reversible	50–95%
Engine/Motor		
Combustion engine	Chemical to kinetic	10–50%
Electric motor	Electrical to kinetic	70–99.99% (> 200 W); 50–90% (10–200 W); 30–60% (< 10 W)
Turbofan	Chemical to kinetic	20-40%
Natural process		
Photosynthesis	Radiative to chemical	up to 6%
Muscle	Chemical to kinetic	14–27%
Appliance		
Household refrigerator	Electrical to thermal	low-end systems ~ 20%; high-end systems ~ 40–50%
Incandescent light bulb	Electrical to radiative	0.7–5.1%, 5–10%
Light-emitting diode (LED)	Electrical to radiative	4.2–53%

Fluorescent lamp	Electrical to radiative	8.0–15.6%, 28%
Low-pressure sodium lamp	Electrical to radiative	15.0–29.0%, 40.5%
Metal-halide lamp	Electrical to radiative	9.5–17.0%, 24%
Switched-mode power supply	Electrical to electrical	currently up to 96% practically
Electric shower	Electrical to thermal	90–95% (multiply with the energy efficiency of electricity generation for comparison with other water-heating systems)
Electric heater	Electrical to thermal	~100% (essentially all energy is converted into heat, multiply with the energy efficiency of electricity generation for comparison with other heating systems)
Others		
Firearm	Chemical to kinetic	~30% (.300 Hawk ammunition)
Electrolysis of water	Electrical to chemical	50–70% (80–94% theoretical maximum)

Direct Energy Conversion

Direct energy conversion (DEC) or simply direct conversion converts a charged particle's kinetic energy into a voltage. It is a scheme for power extraction from nuclear fusion.

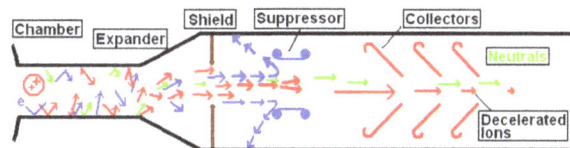

A basic direct converter

History and Theoretical Underpinnings

Electrostatic Direct Collectors

In the middle of the 1960s direct energy conversion was proposed as a method for capturing the energy from the exhaust gas in a fusion reactor. This would generate a direct current of electricity. Richard F. Post at the Lawrence Livermore National Laboratory was an early proponent of the idea. Post reasoned that capturing the energy would require five steps: (1) Ordering the charged particles into linear beam. (2) Separation of positives and negatives. (3) Separating the ions into groups, by their energy. (4) Gathering these ions as they touch collectors. (5) Using these collectors as the positive side in a circuit. Post argued that the efficiency was theoretically determined by the number of collectors.

The Venetian Blind

Designs in the early 1970s by William Barr and Ralph Moir used metal ribbons at an angle to collect these ions. This was called the Venetian Blind design, because the ribbons look like window blinds. Those metal ribbon-like surfaces are more transparent to ions going forward than to ions going backward. Ions pass through surfaces of successively increasing potential until they turn and start back, along a parabolic trajectory. They then see opaque surfaces and are caught. Thus ions are sorted by energy with high-energy ions being caught on high-potential electrodes.

William Barr and Ralph Moir then ran a group which did a series of direct energy conversion experiments through the late 1970s and early 1980s. The first experiments used beams of positives and negatives as fuel, and demonstrated energy capture at a peak efficiency of 65 percent and a minimum efficiency of 50 percent. The following experiments involved a true plasma direct converter that was tested on the Tandem Mirror Experiment (TMX), an operating magnetic mirror fusion reactor. In the experiment, the plasma moved along diverging field lines, spreading it out and converting it into a forward moving beam with a Debye length of a few centimeters. Suppressor grids then reflect the electrons, and collector anodes recovered the ion energy by slowing them down and collecting them at high-potential plates. This machine demonstrated an energy capture efficiency of 48 percent. However, Marshall Rosenbluth argued that keeping the plasma's neutral charge over the very short Debye length distance would be very challenging in practice, though he said that this problem would not occur in every version of this technology.

The Venetian Blind converter can operate with 100 to 150 keV D-T plasma, with an efficiency of about 60% under conditions compatible with economics, and an upper technical conversion efficiency up to 70% ignoring economic limitations.

Periodic Electrostatic Focusing

A second type of electrostatic converter initially proposed by Post, then developed by Barr and Moir, is the Periodic Electrostatic Focusing concept. Like the Venetian Blind concept, it is also a direct collector, but the collector plates are disposed in many stages along the longitudinal axis of an electrostatic focusing channel. As each ion is decelerated along the channel toward zero energy, the particle becomes "over-focused" and is deflected sideways from the beam, then collected. The Periodic Electrostatic Focusing converter typically operates with a 600 keV D-T plasma (as low as 400 keV and up to 800 keV) with efficiency of about 60% under conditions compatible with economics, and an upper technical conversion efficiency up to 90% ignoring economic limitations.

Magnetic Compression-expansion Converter

In addition to electrostatic converters, magnetic converters have also been proposed by Lev Artsimovich in 1963, then Alan Frederic Haught and his team from United Aircraft Research Laboratories in 1970, and Ralph Moir in 1977.

The magnetic direct energy converter is analogous to the internal combustion engine. As the hot plasma expands against a magnetic field, in a manner similar to hot gases expanding against a piston, part of the energy of the internal plasma is inductively converted to an electromagnetic coil, as an EMF (voltage) in the conductor.

This scheme is best used with pulsed devices, because the converter then works like a "magnetic four-stroke engine":

1. Compression: A column of plasma is compressed by a magnetic field that acts like a piston.

2. Thermonuclear burn: The compression heats the plasma to the thermonuclear ignition temperature.

3. Expansion/Power: The expansion of fusion reaction products (charged particles) increases the plasma pressure and pushes the magnetic field outward. A voltage is induced and collected in the electromagnetic coil.

4. Exhaust/Refuel: After expansion, the partially burned fuel is flushed out, and new fuel in the form of gas is introduced and ionized; and the cycle starts again.

In 1973, a team from Los Alamos and Argonne laboratories stated that the thermodynamic efficiency of the magnetic direct conversion cycle from alpha-particle energy to work is 62%.

Traveling-wave Direct Energy Converter

In 1992, a Japan–U.S. joint-team proposed a novel direct energy conversion system for 14.7 MeV protons produced by D-^{3}He fusion reactions, whose energy is too high for electrostatic converters.

The conversion is based on a Traveling-Wave Direct Energy Converter (TWDEC). A gyrotron converter first guides fusion product ions as a beam into a 10-meter long microwave cavity filled with a 10-tesla magnetic field, where 155 MHz microwaves are generated and converted to a high voltage DC output through rectennas.

The Field-Reversed Configuration reactor ARTEMIS in this study was designed with an efficiency of 75%. The traveling-wave direct converter has a maximum projected efficiency of 90%.

Inverse Cyclotron Converter (ICC)

Original direct converters were designed to extract the energy carried by 100 to 800 keV ions produced by D-T fusion reactions. Those electrostatic converters are not suitable for higher energy product ions above 1 MeV generated by other fusion fuels like the D-^{3}He or the p-^{11}B aneutronic fusion reactions.

A much shorter device than the Traveling-Wave Direct Energy Converter has been proposed in 1997 and patented by Tri Alpha Energy, Inc. as an Inverse Cyclotron Converter (ICC).

The ICC is able to decelerate the incoming ions based on experiments made in 1950 by Felix Bloch and Carson D. Jeffries, in order to extract their kinetic energy. The converter operates at 5 MHz and requires a magnetic field of only 0.6 tesla. The linear motion of fusion product ions is converted to circular motion by a magnetic cusp. Energy is collected from the charged particles as they spiral past quadrupole electrodes. More classical electrostatic collectors would also be used for particles with energy less than 1 MeV. The Inverse Cyclotron Converter has a maximum projected efficiency of 90%.

X-ray Photoelectric Converter

A significant amount of the energy released by fusion reactions is composed of electromagnetic radiations, essentially X-rays due to Bremsstrahlung. Those X-rays can not be converted into electric power with the various electrostatic and magnetic direct energy converters listed above, and their energy is lost.

Whereas more classical thermal conversion has been considered with the use of a radiation/boiler/ energy exchanger where the X-ray energy is absorbed by a working fluid at temperatures of several thousand degrees, more recent research done by companies developing nuclear aneutronic fusion reactors, like Lawrenceville Plasma Physics (LPP) with the Dense Plasma Focus, and Tri Alpha Energy, Inc. with the Colliding Beam Fusion Reactor (CBFR), plan to harness the photoelectric and Auger effects to recover energy carried by X-rays and other high-energy photons. Those photoelectric converters are composed of X-ray absorber and electron collector sheets nested concentrically in an onion-like array. Indeed, since X-rays can go through far greater thickness of material than electrons can, many layers are needed to absorb most of the X-rays. LPP announces an overall efficiency of 81% for the photoelectric conversion scheme.

Direct Energy Conversion from Fission Products

In the early 2000s, research was undertaken by Sandia National Laboratories, Los Alamos National Laboratory, The University of Florida, Texas A&M University and General Atomics to use direct conversion to extract energy from fission reactions, essentially, attempting to extract energy from the linear motion of charged particles coming off a fission reaction.

Consolidated Power Generation

Consolidated Power Generation occurs when the total number of energy changing devices, such as automobile engines, decreases, while the total amount of usable energy increases or stays the same.

The consolidation of hydrocarbon power generation to electrical power, and increased output of electricity, must take place for this change to occur. Currently, the prime example of this metamorphosis is the shift from hydrocarbon powered combustion engines to electricity.

A requirement for consolidation is the acceptance of a "common" form of energy, such as electricity, to be the predominant one. The prime environment for the consolidation of power generation is the replacement of combustion engines with electric motors in vehicles.

Even if the electricity was produced in coal plants or by burning crude oil, this would consolidate hydrocarbon power generation from millions of small engines to only a few massive power plants. In doing so the release of carbon would also be consolidated, which would allow for more regulation, and it would also increase demand for technologies that increase efficiency and decrease the release of anything harmful to the environment. This would be a gradual and realistic step.

Advantages of Consolidation

- Large power plants generally have a higher thermal efficiency than small ones

- A few large power plants can be monitored and regulated more easily than many small ones

Disadvantages of Consolidation

- Centralization of electricity generation leads to transmission losses, which will partially offset the increased efficiency of generation

- There may be objections from local residents to the construction of large power plants

- It goes against the Green movement's philosophy Small is Beautiful

- Combined heat and power schemes can be arranged more easily with small power plants

Wave Power

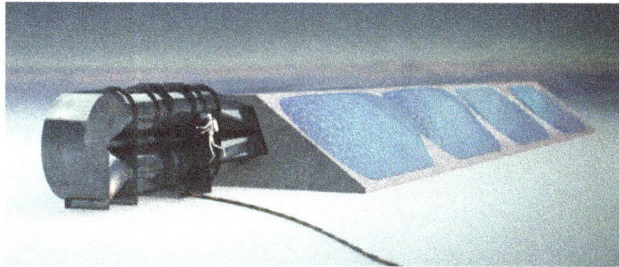

The mWave converter by Bombora Wave Power

SINN Power Wave Energy Converter

Wave power is the transport of energy by wind waves, and the capture of that energy to do useful work – for example, electricity generation, water desalination, or the pumping of water (into reservoirs). A machine able to exploit wave power is generally known as a wave energy converter (WEC).

Wave power is distinct from the diurnal flux of tidal power and the steady gyre of ocean currents. Wave-power generation is not currently a widely employed commercial technology, although there have been attempts to use it since at least 1890. In 2008, the first experimental wave farm was opened in Portugal, at the Aguçadoura Wave Park.

Physical concepts

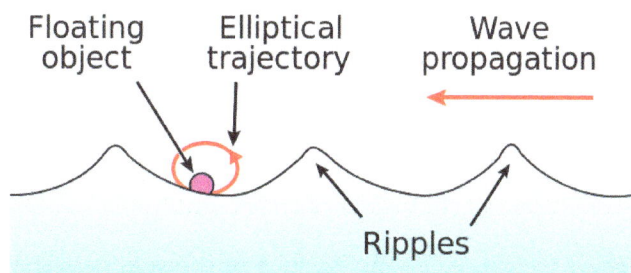

When an object bobs up and down on a ripple in a pond, it follows approximately an elliptical trajectory.

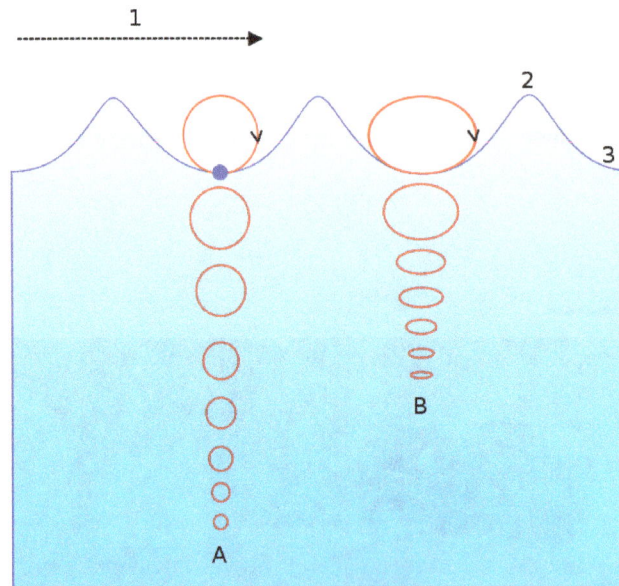

Motion of a particle in an ocean wave A = At deep water.
The elliptical motion of fluid particles decreases rapidly with increasing
depth below the surface. B = At shallow water (ocean floor is now at B).
The elliptical movement of a fluid particle flattens with decreasing depth.
1 = Propagation direction.
2 = Wave crest.
3 = Wave trough.

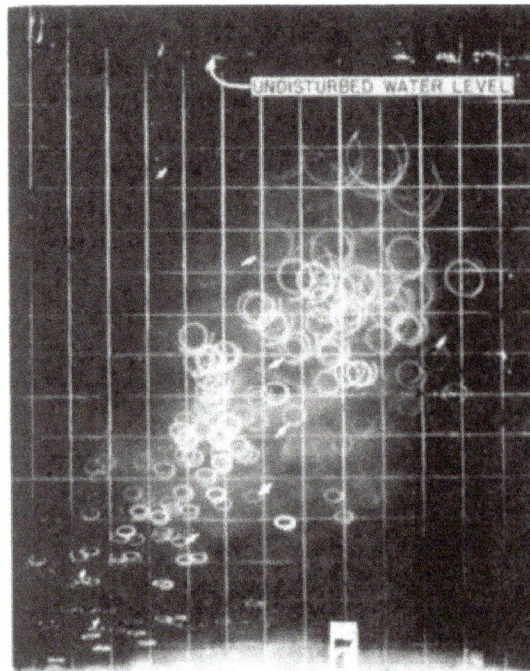

Photograph of the elliptical trajectories of water particles under a – progressive and
periodic – surface gravity wave in a wave flume. The wave conditions are: mean water depth
d = 2.50 ft (0.76 m), wave height H = 0.339 ft (0.103 m), wavelength λ = 6.42 ft (1.96 m), period T = 1.12 s.

Waves are generated by wind passing over the surface of the sea. As long as the waves propagate
slower than the wind speed just above the waves, there is an energy transfer from the wind to the

waves. Both air pressure differences between the upwind and the lee side of a wave crest, as well as friction on the water surface by the wind, making the water to go into the shear stress causes the growth of the waves.

Wave height is determined by wind speed, the duration of time the wind has been blowing, fetch (the distance over which the wind excites the waves) and by the depth and topography of the sea-floor (which can focus or disperse the energy of the waves). A given wind speed has a matching practical limit over which time or distance will not produce larger waves. When this limit has been reached the sea is said to be "fully developed".

In general, larger waves are more powerful but wave power is also determined by wave speed, wavelength, and water density.

Oscillatory motion is highest at the surface and diminishes exponentially with depth. However, for standing waves (clapotis) near a reflecting coast, wave energy is also present as pressure oscillations at great depth, producing microseisms. These pressure fluctuations at greater depth are too small to be interesting from the point of view of wave power.

The waves propagate on the ocean surface, and the wave energy is also transported horizontally with the group velocity. The mean transport rate of the wave energy through a vertical plane of unit width, parallel to a wave crest, is called the wave energy flux (or wave power, which must not be confused with the actual power generated by a wave power device).

Wave Power Formula

In deep water where the water depth is larger than half the wavelength, the wave energy flux is

$$P = \frac{\rho g^2}{64\pi} H_{m0}^2 T_e \approx \left(0.5 \frac{kW}{m^3 \cdot s}\right) H_{m0}^2 T_e,$$

with P the wave energy flux per unit of wave-crest length, H_{m0} the significant wave height, T_e the wave energy period, ρ the water density and g the acceleration by gravity. The above formula states that wave power is proportional to the wave energy period and to the square of the wave height. When the significant wave height is given in metres, and the wave period in seconds, the result is the wave power in kilowatts (kW) per metre of wavefront length.

Example: Consider moderate ocean swells, in deep water, a few km off a coastline, with a wave height of 3 m and a wave energy period of 8 seconds. Using the formula to solve for power, we get

$$P \approx 0.5 \frac{kW}{m^3 \cdot s} (3 \cdot m)^2 (8 \cdot s) \approx 36 \frac{kW}{m},$$

meaning there are 36 kilowatts of power potential per meter of wave crest.

In major storms, the largest waves offshore are about 15 meters high and have a period of about 15 seconds. According to the above formula, such waves carry about 1.7 MW of power across each metre of wavefront.

An effective wave power device captures as much as possible of the wave energy flux. As a result, the waves will be of lower height in the region behind the wave power device.

Wave Energy and Wave-energy Flux

In a sea state, the average(mean) energy density per unit area of gravity waves on the water surface is proportional to the wave height squared, according to linear wave theory:

$$E = \frac{1}{16} \rho g H_{m0}^2,$$

where E is the mean wave energy density per unit horizontal area (J/m²), the sum of kinetic and potential energy density per unit horizontal area. The potential energy density is equal to the kinetic energy, both contributing half to the wave energy density E, as can be expected from the equipartition theorem. In ocean waves, surface tension effects are negligible for wavelengths above a few decimetres.

As the waves propagate, their energy is transported. The energy transport velocity is the group velocity. As a result, the wave energy flux, through a vertical plane of unit width perpendicular to the wave propagation direction, is equal to:

$$P = Ec_g,$$

with c_g the group velocity (m/s). Due to the dispersion relation for water waves under the action of gravity, the group velocity depends on the wavelength λ, or equivalently, on the wave period T. Further, the dispersion relation is a function of the water depth h. As a result, the group velocity behaves differently in the limits of deep and shallow water, and at intermediate depths:

Properties of gravity waves on the surface of deep water, shallow water and at intermediate depth, according to linear wave theory					
quantity	symbol	units	deep water ($h > \frac{1}{2}\lambda$)	shallow water ($h < 0.05\lambda$)	intermediate depth (all λ and h)
phase velocity	$c_p = \dfrac{\lambda}{T} = \dfrac{\omega}{k}$	m / s	$\dfrac{g}{2\pi}T$	$\sqrt{gh}$	$\sqrt{\dfrac{g\lambda}{2\pi}\tanh\left(\dfrac{2\pi h}{\lambda}\right)}$
group velocity	$c_g = c_p^2\dfrac{\partial(\lambda/c_p)}{\partial\lambda} = \dfrac{\partial\omega}{\partial k}$	m / s	$\dfrac{g}{4\pi}T$	$\sqrt{gh}$	$\dfrac{1}{2}c_p\left(1 + \dfrac{4\pi h}{\lambda}\dfrac{1}{\sinh\left(\dfrac{4\pi h}{\lambda}\right)}\right)$

ratio	$\dfrac{c_g}{c_p}$	–	$\dfrac{1}{2}$	1	$\dfrac{1}{2}\left(1+\dfrac{4\pi h}{\lambda}\dfrac{1}{\sinh\left(\dfrac{4\pi h}{\lambda}\right)}\right)$
wavelength	λ	m	$\dfrac{g}{2\pi}T^2$	$T\sqrt{gh}$	for given period T, the solution of: $\left(\dfrac{2\pi}{T}\right)^2=\dfrac{2\pi g}{\lambda}\tanh\left(\dfrac{2\pi h}{\lambda}\right)$
general					
wave energy density	E	J / m²	$\dfrac{1}{16}\rho g H_{m0}^2$		
wave energy flux	P	W / m	$E\,c_g$		
angular frequency	ω	rad / s	$\dfrac{2\pi}{T}$		
wavenumber	k	rad / m	$\dfrac{2\pi}{\lambda}$		

Deep-water Characteristics and Opportunities

Deep water corresponds with a water depth larger than half the wavelength, which is the common situation in the sea and ocean. In deep water, longer-period waves propagate faster and transport their energy faster. The deep-water group velocity is half the phase velocity. In shallow water, for wavelengths larger than about twenty times the water depth, as found quite often near the coast, the group velocity is equal to the phase velocity.

History

The first known patent to use energy from ocean waves dates back to 1799, and was filed in Paris by Girard and his son. An early application of wave power was a device constructed around 1910 by Bochaux-Praceique to light and power his house at Royan, near Bordeaux in France. It appears that this was the first oscillating water-column type of wave-energy device. From 1855 to 1973 there were already 340 patents filed in the UK alone.

Modern scientific pursuit of wave energy was pioneered by Yoshio Masuda's experiments in the 1940s. He tested various concepts of wave-energy devices at sea, with several hundred units used to power navigation lights. Among these was the concept of extracting power from the angular motion at the joints of an articulated raft, which was proposed in the 1950s by Masuda.

A renewed interest in wave energy was motivated by the oil crisis in 1973. A number of university researchers re-examined the potential to generate energy from ocean waves, among whom notably were Stephen Salter from the University of Edinburgh, Kjell Budal and Johannes Falnes from Norwegian Institute of Technology (now merged into Norwegian University of Science and Technology), Michael E. McCormick from U.S. Naval Academy, David Evans from Bristol University, Michael French from University of Lancaster, Nick Newman and C. C. Mei from MIT.

Stephen Salter's 1974 invention became known as Salter's duck or *nodding duck*, although it was officially referred to as the Edinburgh Duck. In small scale controlled tests, the Duck's curved cam-like body can stop 90% of wave motion and can convert 90% of that to electricity giving 81% efficiency.

In the 1980s, as the oil price went down, wave-energy funding was drastically reduced. Nevertheless, a few first-generation prototypes were tested at sea. More recently, following the issue of climate change, there is again a growing interest worldwide for renewable energy, including wave energy.

The world's first marine energy test facility was established in 2003 to kick start the development of a wave and tidal energy industry in the UK. Based in Orkney, Scotland, the European Marine Energy Centre (EMEC) has supported the deployment of more wave and tidal energy devices than at any other single site in the world. EMEC provides a variety of test sites in real sea conditions. Its grid-connected wave test site is situated at Billia Croo, on the western edge of the Orkney mainland, and is subject to the full force of the Atlantic Ocean with seas as high as 19 metres recorded at the site. Wave energy developers currently testing at the centre include Aquamarine Power, Pelamis Wave Power, ScottishPower Renewables and Wello.

Modern Technology

Wave power devices are generally categorized by the method used to capture the energy of the waves, by location and by the power take-off system. Locations are shoreline, nearshore and offshore. Types of power take-off include: hydraulic ram, elastomeric hose pump, pump-to-shore, hydroelectric turbine, air turbine, and linear electrical generator. When evaluating wave energy as a technology type, it is important to distinguish between the four most common approaches: point absorber buoys, surface attenuators, oscillating water columns, and overtopping devices.

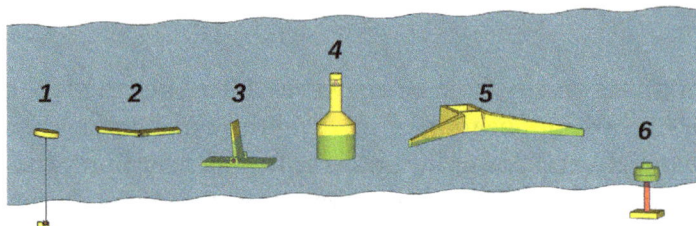

Generic wave energy concepts: 1. Point absorber, 2. Attenuator, 3. Oscillating wave surge converter, 4. Oscillating water column, 5. Overtopping device, 6. Submerged pressure differential

Point Absorber Buoy

This device floats on the surface of the water, held in place by cables connected to the seabed. Buoys use the rise and fall of swells to drive hydraulic pumps and generate electricity. EMF generated by electrical transmission cables and acoustics of these devices may be a concern for marine organisms. The presence of the buoys may affect fish, marine mammals, and birds as potential minor collision risk and roosting sites. Potential also exists for entanglement in mooring lines. Energy removed from the waves may also affect the shoreline, resulting in a recommendation that sites remain a considerable distance from the shore.

Surface Attenuator

These devices act similarly to point absorber buoys, with multiple floating segments connected to one another and are oriented perpendicular to incoming waves. A flexing motion is created by swells that drive hydraulic pumps to generate electricity. Environmental effects are similar to those of point absorber buoys, with an additional concern that organisms could be pinched in the joints.

Oscillating Wave Surge Converter

These devices typically have one end fixed to a structure or the seabed while the other end is free to move. Energy is collected from the relative motion of the body compared to the fixed point. Oscillating wave surge converters often come in the form of floats, flaps, or membranes. Environmental concerns include minor risk of collision, artificial reefing near the fixed point, EMF effects from subsea cables, and energy removal effecting sediment transport. Some of these designs incorporate parabolic reflectors as a means of increasing the wave energy at the point of capture. These capture systems use the rise and fall motion of waves to capture energy. Once the wave energy is captured at a wave source, power must be carried to the point of use or to a connection to the electrical grid by transmission power cables.

Oscillating Water Column

Oscillating Water Column devices can be located on shore or in deeper waters offshore. With an air chamber integrated into the device, swells compress air in the chambers forcing air through an air turbine to create electricity. Significant noise is produced as air is pushed through the turbines, potentially affecting birds and other marine organisms within the vicinity of the device. There is also concern about marine organisms getting trapped or entangled within the air chambers.

Overtopping Device

Overtopping devices are long structures that use wave velocity to fill a reservoir to a greater water level than the surrounding ocean. The potential energy in the reservoir height is then captured with low-head turbines. Devices can be either on shore or floating offshore. Floating devices will have environmental concerns about the mooring system affecting benthic organisms, organisms becoming entangled, or EMF effects produced from subsea cables. There is also some concern regarding low levels of turbine noise and wave energy removal affecting the nearfield habitat.

Submerged Pressure Differential

Submerged pressure differential based converters are a comparatively newer technology utilizing flexible (usually reinforced rubber) membranes to extract wave energy. These converters use the difference in pressure at different locations below a wave to produce a pressure difference within a closed power take-off fluid system. This pressure difference is usually used to produce flow, which drives a turbine and electrical generator. Submerged pressure differential converters frequently use flexible membranes as the working surface between the ocean and the power take-off system. Membranes offer the advantage over rigid structures of being compliant and low mass, which can produce more direct coupling with the wave's energy. Their compliant nature also allows for large changes in the geometry of the working surface, which can be used to tune the response of the converter for specific wave conditions and to protect it from excessive loads in extreme conditions.

A submerged converter may be positioned either on the sea floor or in midwater. In both cases, the converter is protected from water impact loads which can occur at the free surface. Wave loads also diminish in non-linear proportion to the distance below the free surface. This means that by optimizing the depth of submergence for such a converter, a compromise between protection from extreme loads and access to wave energy can be found. Submerged WECs also have the potential to reduce the impact on marine amenity and navigation, as they are not at the surface. Example's of submerged pressure differential converters include M3 Wave and Bombora Wave Power's mWave.

Environmental Effects

Common environmental concerns associated with marine energy developments include:

- The risk of marine mammals and fish being struck by tidal turbine blades;

- The effects of EMF and underwater noise emitted from operating marine energy devices;

- The physical presence of marine energy projects and their potential to alter the behavior of marine mammals, fish, and seabirds with attraction or avoidance;

- The potential effect on nearfield and farfield marine environment and processes such as sediment transport and water quality.

The Tethys database provides access to scientific literature and general information on the potential environmental effects of wave energy.

Potential

The worldwide resource of wave energy has been estimated to be greater than 2 TW. Locations with the most potential for wave power include the western seaboard of Europe, the northern coast of the UK, and the Pacific coastlines of North and South America, Southern Africa, Australia, and New Zealand. The north and south temperate zones have the best sites for capturing wave power. The prevailing westerlies in these zones blow strongest in winter.

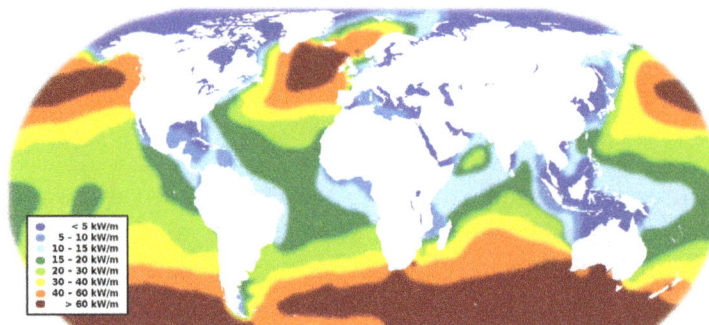

World wave energy resource map

Challenges

There is a potential impact on the marine environment. Noise pollution, for example, could have negative impact if not monitored, although the noise and visible impact of each design vary greatly. Other biophysical impacts (flora and fauna, sediment regimes and water column structure and flows) of scaling up the technology is being studied. In terms of socio-economic challenges, wave farms can result in the displacement of commercial and recreational fishermen from productive fishing grounds, can change the pattern of beach sand nourishment, and may represent hazards to safe navigation. Waves generate about 2,700 gigawatts of power. Of those 2,700 gigawatts, only about 500 gigawatts can be captured with the current technology.

Portugal

- The Aguçadoura Wave Farm was the world's first wave farm. It was located 5 km (3 mi) offshore near Póvoa de Varzim, north of Porto, Portugal. The farm was designed to use three Pelamis wave energy converters to convert the motion of the ocean surface waves into electricity, totalling to 2.25 MW in total installed capacity. The farm first generated electricity in July 2008 and was officially opened on September 23, 2008, by the Portuguese Minister of Economy. The wave farm was shut down two months after the official opening in November 2008 as a result of the financial collapse of Babcock & Brown due to the global economic crisis. The machines were off-site at this time due to technical problems, and although resolved have not returned to site and were subsequently scrapped in 2011 as the technology had moved on to the P2 variant as supplied to Eon and Scottish Power Renewables. A second phase of the project planned to increase the installed capacity to 21 MW using a further 25 Pelamis machines is in doubt following Babcock's financial collapse.

United Kingdom

- Funding for a 3 MW wave farm in Scotland was announced on February 20, 2007, by the Scottish Executive, at a cost of over 4 million pounds, as part of a £13 million funding package for marine power in Scotland. The first machine was launched in May 2010.

- A facility known as Wave hub has been constructed off the north coast of Cornwall, England, to facilitate wave energy development. The Wave hub will act as giant extension cable, allowing arrays of wave energy generating devices to be connected to the electricity grid. The Wave hub will initially allow 20 MW of capacity to be connected, with potential expansion to 40 MW. Four device manufacturers have so far expressed interest in connecting to the Wave hub. The scientists have calculated that wave energy gathered at Wave Hub will be enough to power up to 7,500 households. The site has the potential to save greenhouse gas emissions of about 300,000 tons of carbon dioxide in the next 25 years.

Australia

- Bombora Wave Power (Bombora Wave Power Pty Ltd) is based in Perth, Western Australia and is currently developing the *mWave* flexible membrane converter. Bombora is currently preparing for a commercial pilot project in Peniche, Portugal.

- A CETO wave farm off the coast of Western Australia has been operating to prove commercial viability and, after preliminary environmental approval, underwent further development. In early 2015 a $100 million, multi megawatt system was connected to the grid, with all the electricity being bought to power HMAS Stirling naval base. Two fully submerged buoys which are anchored to the seabed, transmit the energy from the ocean swell through hydraulic pressure onshore; to drive a generator for electricity, and also to produce fresh water. As of 2015 a third buoy is planned for installation.

- Ocean Power Technologies (OPT Australasia Pty Ltd) is developing a wave farm connected to the grid near Portland, Victoria through a 19 MW wave power station. The project has received an AU $66.46 million grant from the Federal Government of Australia.

- Oceanlinx will deploy a commercial scale demonstrator off the coast of South Australia at Port MacDonnell before the end of 2013. This device, the *greenWAVE*, has a rated electrical capacity of 1MW. This project has been supported by ARENA through the Emerging Renewables Program. The *greenWAVE* device is a bottom standing gravity structure, that does not require anchoring or seabed preparation and with no moving parts below the surface of the water.

United States

- Reedsport, Oregon – a commercial wave park on the west coast of the United States located 2.5 miles offshore near Reedsport, Oregon. The first phase of this project is for ten PB150 PowerBuoys, or 1.5 megawatts. The Reedsport wave farm was scheduled for installation spring 2013. In 2013, the project had ground to a halt because of legal and technical problems.

- Kaneohe Bay Oahu, Hawaii - Navy's Wave Energy Test Site (WETS) currently testing the Azura wave power device The Azura wave power device is 45-ton wave energy converter located at a depth of 30 meters in Kaneohe Bay.

Patents

- WIPO patent application WO2016032360 — 2016 *Pumped-storage system* – "Pressure buffering hydro power" patent application

- U.S. Patent 8,806,865 — 2011 *Ocean wave energy harnessing device* – Pelamis/Salter's Duck Hybrid patent

- U.S. Patent 3,928,967 — 1974 *Apparatus and method of extracting wave energy* – The original "Salter's Duck" patent

- U.S. Patent 4,134,023 — 1977 *Apparatus for use in the extraction of energy from waves on water* – Salter's method for improving "duck" efficiency

- U.S. Patent 6,194,815 — 1999 *Piezoelectric rotary electrical energy generator*

- Wave energy converters utilizing pressure differences US 20040217597 A1 — 2004 *Wave energy converters utilizing pressure differences*

References

- Graham, Karen." First wave-produced power in U.S. goes online in Hawaii" Digital Journal. 19 September 2016. Web Accessed 22 September 2016

- Moir, Ralph W. (April 1977). "Chapter 5: Direct Energy Conversion in Fusion Reactors". In Considine, Douglas M. Energy Technology Handbook (PDF). NY: McGraw-Hill. pp. 150–154. ISBN 978-0070124301

- "All in 1 LED Lighting Solutions Guide". PhilipsLumileds.com. Philips. 2012-10-04. p. 15. Archived from the original (PDF) on 2013-03-31. Retrieved 2015-11-18

- Falnes, J. (2007). "A review of wave-energy extraction". Marine Structures. 20 (4): 185–201. doi:10.1016/j.marstruc.2007.09.001

- Morris-Thomas; Irvin, Rohan J.; Thiagarajan, Krish P.; et al. (2007). "An Investigation Into the Hydrodynamic Efficiency of an Oscillating Water Column". Journal of Offshore Mechanics and Arctic Engineering. 129 (4): 273–278. doi:10.1115/1.2426992

- Tucker, M.J.; Pitt, E.G. (2001). "2". In Bhattacharyya, R.; McCormick, M.E. Waves in ocean engineering (1st ed.). Oxford: Elsevier. pp. 35–36. ISBN 0080435661

- Farley, F. J. M. & Rainey, R. C. T. (2006). "Radical design options for wave-profiling wave energy converters" (PDF). International Workshop on Water Waves and Floating Bodies. Loughborough. Retrieved 2009-12-18

- Bloch, F.; Jeffries, C. (1950). "A Direct Determination of the Magnetic Moment of the Proton in Nuclear Mag-netons". Physical Review. 80 (2): 305. Bibcode:1950PhRv...80..305B. doi:10.1103/PhysRev.80.305

- Barr, W. L.; Moir, R. W.; Hamilton, G. W. (1982). "Experimental results from a beam direct converter at 100 kV". Journal of Fusion Energy. 2 (2): 131. Bibcode:1982JFuE....2..131B. doi:10.1007/BF01054580

- R. G. Dean & R. A. Dalrymple (1991). Water wave mechanics for engineers and scientists. Advanced Series on Ocean Engineering. 2. World Scientific, Singapore. ISBN 978-981-02-0420-4

- Fyall, Jenny (May 19, 2010). "600ft 'sea snake' to harness power of Scotland". The Scotsman. Edinburgh. pp. 10–11. Retrieved 2010-05-19

- Moir, R. W.; Barr, W. L. (1973). ""Venetian-blind" direct energy converter for fusion reactors" (PDF). Nuclear Fusion. 13: 35. doi:10.1088/0029-5515/13/1/005

- Moir, R. W.; Barr, W. L.; Miley, G. H. (1974). "Surface requirements for electrostatic direct energy converters" (PDF). Journal of Nuclear Materials. 53: 86. Bibcode:1974JNuM...53...86M. doi:10.1016/0022-3115(74)90225-6

- "WA wave energy project turned on to power naval base at Garden Island". ABC News Online. Australian Broadcasting Corporation. 18 February 2015. Retrieved 20 February 2015

- l.c. Brown (2002). "Direct Energy Conversion Fission Reactor Annual Report for the Period August 15,2000 Through September 30,2001". doi:10.2172/805252

Permissions

All chapters in this book are published with permission under the Creative Commons Attribution Share Alike License or equivalent. Every chapter published in this book has been scrutinized by our experts. Their significance has been extensively debated. The topics covered herein carry significant information for a comprehensive understanding. They may even be implemented as practical applications or may be referred to as a beginning point for further studies.

We would like to thank the editorial team for lending their expertise to make the book truly unique. They have played a crucial role in the development of this book. Without their invaluable contributions this book wouldn't have been possible. They have made vital efforts to compile up to date information on the varied aspects of this subject to make this book a valuable addition to the collection of many professionals and students.

This book was conceptualized with the vision of imparting up-to-date and integrated information in this field. To ensure the same, a matchless editorial board was set up. Every individual on the board went through rigorous rounds of assessment to prove their worth. After which they invested a large part of their time researching and compiling the most relevant data for our readers.

The editorial board has been involved in producing this book since its inception. They have spent rigorous hours researching and exploring the diverse topics which have resulted in the successful publishing of this book. They have passed on their knowledge of decades through this book. To expedite this challenging task, the publisher supported the team at every step. A small team of assistant editors was also appointed to further simplify the editing procedure and attain best results for the readers.

Apart from the editorial board, the designing team has also invested a significant amount of their time in understanding the subject and creating the most relevant covers. They scrutinized every image to scout for the most suitable representation of the subject and create an appropriate cover for the book.

The publishing team has been an ardent support to the editorial, designing and production team. Their endless efforts to recruit the best for this project, has resulted in the accomplishment of this book. They are a veteran in the field of academics and their pool of knowledge is as vast as their experience in printing. Their expertise and guidance has proved useful at every step. Their uncompromising quality standards have made this book an exceptional effort. Their encouragement from time to time has been an inspiration for everyone.

The publisher and the editorial board hope that this book will prove to be a valuable piece of knowledge for students, practitioners and scholars across the globe.

Index

www.ingramcontent.com/pod-product-compliance
Lightning Source LLC
Chambersburg PA
CBHW082034190326
41458CB00010B/3366